MATH *for* WELDERS

Sixth Edition

NINO MARION

G-W
PUBLISHER

Designed to help today's welding students learn and apply basic math concepts that will serve them throughout their careers as welders.

G-W PUBLISHER EduHub

Be Digital Ready on Day One with EduHub

EduHub provides a solid base of knowledge and instruction for digital and blended classrooms. This easy-to-use learning hub delivers the foundation and tools that improve student retention and facilitate instructor efficiency. For the student, EduHub offers an online collection of eBook content, interactive practice, and test preparation. Additionally, students have the ability to view and submit assessments, track personal performance, and view feedback via the Student Report option. For instructors, EduHub provides a turnkey, fully integrated solution with course management tools to deliver content, assessments, and feedback to students quickly and efficiently. The integrated approach results in improved student outcomes and instructor flexibility.

Flamingo Images/Shutterstock.com

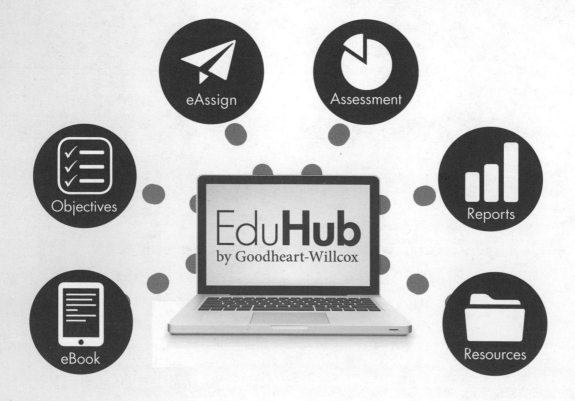

eAssign

Assessment

Objectives

Reports

EduHub
by Goodheart-Willcox

eBook

Resources

eBook

The EduHub eBook engages students by providing the ability to take notes and highlight key concepts to remember.

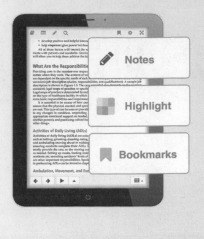

Objectives

Course objectives at the beginning of each eBook chapter help students stay focused and provide benchmarks for instructors to evaluate student progress.

eAssign

eAssign makes it easy for instructors to assign, deliver, and assess student engagement. Coursework can be administered to individual students or the entire class.

goodluz/Shutterstock.com

Assessment

Self-assessment opportunities enable students to gauge their understanding as they progress through the course. In addition, formative assessment tools for instructor use provide efficient evaluation of student mastery of content.

Reports

Reports, for both students and instructors, provide performance results in an instant. Analytics reveal individual student and class achievements for easy monitoring of success.

	🖨 Print	⬇ Export
Score	**Items**	
100%	●	●
80%	●	●
100%	●	●
80%	●	●
100%	●	●
100%	●	●

Instructor Resources

Instructors will find all the support they need to make preparation and classroom instruction more efficient and easier than ever. Lesson plans and answer keys provide an organized, proven approach to classroom management.

Learn more about EduHub at www.g-w.com/eduhub

GUIDED TOUR

Key Terms

listed at the beginning of each unit are important terms to be learned. The terms are in **_bold/italic color_** type in the text and are defined in the Glossary.

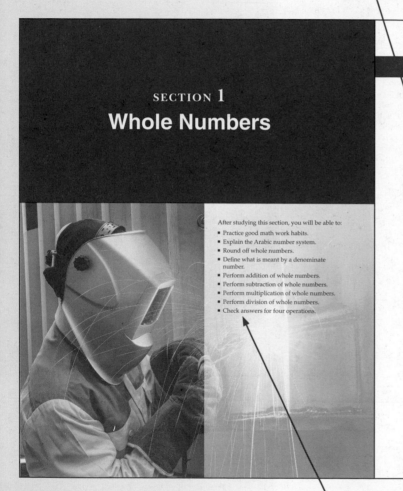

SECTION 1
Whole Numbers

After studying this section, you will be able to:

- Practice good math work habits.
- Explain the Arabic number system.
- Round off whole numbers.
- Define what is meant by a denominate number.
- Perform addition of whole numbers.
- Perform subtraction of whole numbers.
- Perform multiplication of whole numbers.
- Perform division of whole numbers.
- Check answers for four operations.

UNIT **1**

Introduction to Whole Numbers

anekoho/Shutterstock.com

Key Terms

approximate numbers	decimal number system	place value
Arabic number system	denominate numbers	rounding

Introduction

A mastery of mathematics is one of the skills expected of you as a welder. Fortunately, the method of acquiring math skills is no different than any other skill. You learn the principles and then practice them in repeated applications until they become second nature.

What about the questions "Do I need to know math if I know how to use a calculator?" and "Is it necessary to develop a skill in mental calculation and a mastery of math facts?" There are practical reasons why the answer to these questions is yes. First, your employer and supervisor will expect you to be knowledgeable in math. They will expect that, as a tradesperson, you will bring basic math skills and knowledge of math facts to the job. If you begin your training in math by relying on a calculator, you will do yourself a great disservice. A calculator does not give you the confidence and understanding that comes with building a skill.

Also, when a group of welders is discussing the math aspects of a job (dimensions, weights, volumes, costs, etc.), you must be able to understand and contribute to the discussion. If your math skills are weak, you will likely feel uncomfortable about joining in on this part of your work. Besides, there is certain admiration granted to a tradesperson by fellow workers who recognize that the person is clearly proficient in math.

Math Work Habits

Although math is almost entirely a mental activity, there is a small but important tangible component to it. All your calculations, dimensions, and notes must be readable. Correct results are, of course, the final product of math, but the calculations should be done with a high degree of neatness and care. Orderly work in math is an important factor in arriving at correct answers. The following are some useful suggestions:

- Write your numbers fairly large.
- Write the numbers 4 and 9 carefully. When poorly written, these numbers are easily confused.

Section Objectives

clearly identify the knowledge and skills to be obtained when the units in the section are completed.

Introduction

introduces the subject matter of the unit, guiding the student on their learning path for the unit.

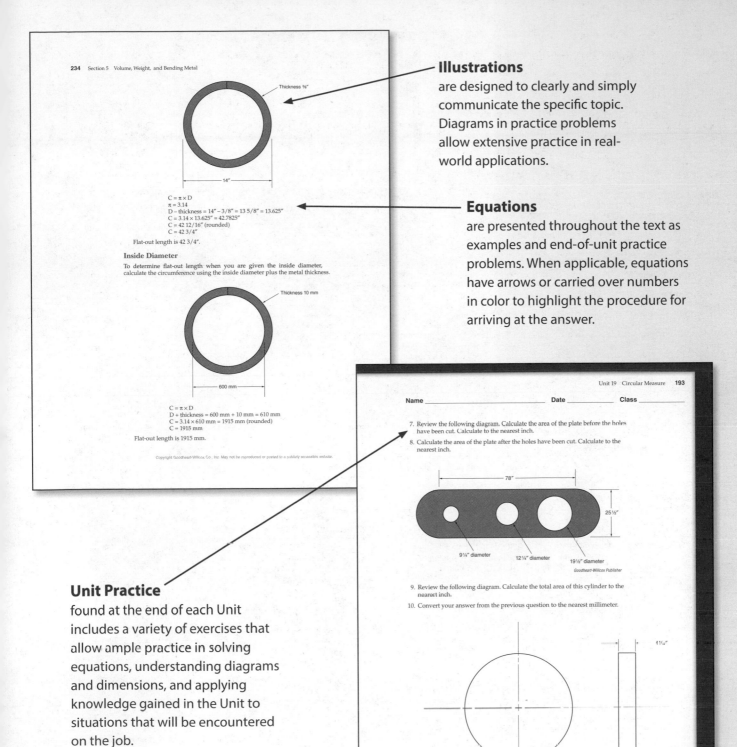

Illustrations

are designed to clearly and simply communicate the specific topic. Diagrams in practice problems allow extensive practice in real-world applications.

Equations

are presented throughout the text as examples and end-of-unit practice problems. When applicable, equations have arrows or carried over numbers in color to highlight the procedure for arriving at the answer.

Unit Practice

found at the end of each Unit includes a variety of exercises that allow ample practice in solving equations, understanding diagrams and dimensions, and applying knowledge gained in the Unit to situations that will be encountered on the job.

Page 234 content (illustration)

234   Section 5   Volume, Weight, and Bending Metal

Thickness ⅜″

14″

$C = \pi \times D$
$\pi = 3.14$
D – thickness = 14″ – 3/8″ = 13 5/8″ = 13.625″
$C = 3.14 \times 13.625″ = 42.7825″$
C = 42 12/16″ (rounded)
C = 42 3/4″

Flat-out length is 42 3/4″.

Inside Diameter

To determine flat-out length when you are given the inside diameter, calculate the circumference using the inside diameter plus the metal thickness.

Thickness 10 mm

600 mm

$C = \pi \times D$
D + thickness = 600 mm + 10 mm = 610 mm
C = 3.14 × 610 mm = 1915 mm (rounded)
C = 1915 mm

Flat-out length is 1915 mm.

Page 193 content (Unit Practice)

Name _____   Date _____   Class _____

7. Review the following diagram. Calculate the area of the plate before the holes have been cut. Calculate to the nearest inch.

8. Calculate the area of the plate after the holes have been cut. Calculate to the nearest inch.

78″

25 ½″

9¼″ diameter   12¼″ diameter   19½″ diameter

Goodheart-Willcox Publisher

9. Review the following diagram. Calculate the total area of this cylinder to the nearest inch.

10. Convert your answer from the previous question to the nearest millimeter.

4¼″

47″ diameter

Goodheart-Willcox Publisher

TOOLS FOR STUDENT AND INSTRUCTOR SUCCESS

EduHub

EduHub provides a solid base of knowledge and instruction for digital and blended classrooms. This easy-to-use learning hub provides the foundation and tools that improve student retention and facilitate instructor efficiency.

For the student, EduHub offers an online collection of eBook content, interactive practice, and test preparation. Additionally, students have the ability to view and submit assessments, track personal performance, and view feedback via the Student Report option. For the instructor, EduHub provides a turnkey, fully integrated solution with course management tools to deliver content, assessments, and feedback to students quickly and efficiently. The integrated approach results in improved student outcomes and instructor flexibility. Be digital ready on day one with EduHub!

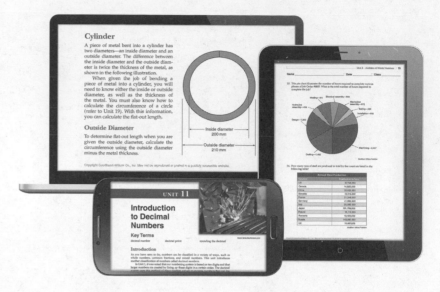

- **eBook content.** EduHub includes the textbook in an online format. The eBook is interactive, with highlighting, magnification, and note-taking features.

- **Assignments.** In EduHub, students can complete the text practice questions as online assignments.

Student Tools

Student Text

Math for Welders is a combination text and workbook designed to help welding students learn and apply basic math skills. The basic concept behind each math operation is explained at the opening of each unit. Next, students are given clear instruction for performing the operation. Each unit includes a variety of welding-related practice problems to reinforce what the students have learned. The practice problems are identical to the types of problems the students will be required to solve in a welding shop. This helps the students develop solid troubleshooting skills that will serve them throughout their careers as welders.

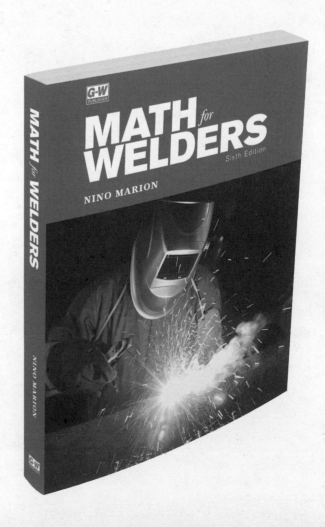

Instructor Tools

LMS Integration

Integrate Goodheart-Willcox content in your Learning Management System for a seamless user experience for both you and your students. Contact your G-W Educational Consultant for ordering information or visit **www.g-w.com/lms-integration.**

Instructor Resources

Instructor Resources provide all the support needed to make preparation and classroom instruction easier than ever. Available in one accessible location, you will find instructor resources and assessment software with question banks. These resources are available as a subscription and can be accessed at school, at home, or on the go.

Instructor Resources One resource provides instructors with time-saving preparation tools such as answer keys and editable lesson plans.

Assessment Software with Question Banks

Administer and manage assessments to meet your classroom needs. The following options are available through the Respondus Test Bank Network:

- A Respondus 4.0 license can be purchased directly from Respondus, which enables you to easily create tests that can be printed on paper or published directly to a variety of Learning Management Systems. Once the question files are published to an LMS, exams can be distributed to students with results reported directly to the LMS gradebook.

- Respondus LE is a limited version of Respondus 4.0 and is free with purchase of the Instructor Resources. It allows you to download question banks and create assessments that can be printed or saved as a paper test.

G-W Integrated Learning Solution

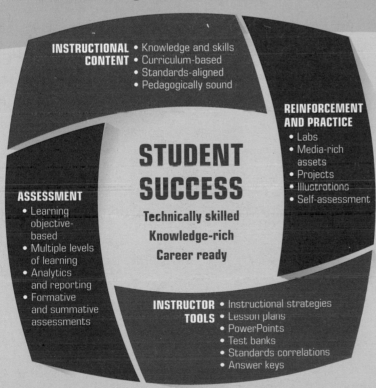

INSTRUCTIONAL CONTENT
- Knowledge and skills
- Curriculum-based
- Standards-aligned
- Pedagogically sound

REINFORCEMENT AND PRACTICE
- Labs
- Media-rich assets
- Projects
- Illustrations
- Self-assessment

ASSESSMENT
- Learning objective-based
- Multiple levels of learning
- Analytics and reporting
- Formative and summative assessments

INSTRUCTOR TOOLS
- Instructional strategies
- Lesson plans
- PowerPoints
- Test banks
- Standards correlations
- Answer keys

STUDENT SUCCESS

Technically skilled

Knowledge-rich

Career ready

The G-W Integrated Learning Solution offers easy-to-use resources that help students and instructors achieve success.

▶ **EXPERT AUTHORS**
▶ **TRUSTED REVIEWERS**
▶ **100 YEARS OF EXPERIENCE**

EMPLOYABILITY SKILLS · TECHNICAL SKILLS · ACADEMIC KNOWLEDGE · INDUSTRY RECOGNIZED STANDARDS

MATH *for* WELDERS

Sixth Edition

NINO MARION
St. Clair College

Publisher
The Goodheart-Willcox Company, Inc.
Tinley Park, Illinois
www.g-w.com

Library of Congress Catalog Card Number 2018038354

ISBN 978-1-63563-658-1

3 4 5 6 7 8 9 – 20 – 23 22 21

Library of Congress Cataloging-in-Publication Data

Names: Marion, Nino, author.
Title: Math for welders / by Nino Marion (St. Clair College).
Description: 6th edition. | Tinley Park, Illinois : The Goodheart-Willcox
  Company, Inc., [2020] | Includes index.
Identifiers: LCCN 2018038354 | ISBN 9781635636581
Subjects: LCSH: Welding--Mathematics.
Classification: LCC TS227.2 .M37 2020 | DDC 671.5/2--dc23 LC record available at
https://lccn.loc.gov/2018038354

Preface

Math for Welders is a combination textbook and workbook that teaches basic mathematics skills and provides practical exercises that are useful in the welding field. The topics are presented in a step-by-step approach. Clear examples facilitate learning and improve understanding of the basics. Students have different math education backgrounds and may have learned their math skills through different approaches. This text teaches traditional basic math in a concise and straightforward way.

This write-in textbook covers the following areas of instruction:

- Whole numbers
- Common fractions
- Decimal numbers
- Measurement
- Volume
- Weight
- Bending metal
- Percentages
- SI metric system

Math for Welders contains practice problems in each unit—over 900 in all! Many welding-related problems are included to sharpen the application of basic mathematics to situations commonly encountered in welding tasks. This text allows instructors to provide realistic, relevant problems that students will find meaningful and interesting to solve.

Space is provided in the book to work the problems and to record answers. Answers to odd-numbered practice problems are located in the back of the book. By referring to these answers, students can check their progress while studying.

Key terms in each unit are defined in the Glossary. This book is designed to provide the student with a precise math vocabulary and a thorough understanding of the meaning of terms used in solving math problems.

Math knowledge and skills are just as important to career development as the ability to maintain welding equipment or to weld in various positions. Math skills are critical to the ability to accurately read blueprints, measure materials and projects, calculate shapes and volume, form accurate joints, and a multitude of other tasks. Learning the skills taught in this book puts students on a path to confidence and success on the job!

About the Author

After graduating from Wayne State University, Mr. Marion invested several years working for Chrysler Canada. He then transitioned to a teaching position at St. Clair College, focusing on math and technical drawing. Mr. Marion used his collective knowledge and experience to author *Math for Welders* as a career capstone.

Reviewers

The author and publisher wish to thank the following industry and teaching professionals for their valuable input into the development of *Math for Welders*.

Charles A. Bartocci
Dabney S. Lancaster Community
  College
Clifton Forge, Virginia

Christy Dvorsky
Front Range Community College
Fort Collins, Colorado

Billy M. Elliott, Jr.
Southern Crescent Technical College
Griffin, Georgia

Carl Johnson
Southwest Applied Technology College
Cedar City, Utah

Darin Owens
Metropolitan Community College
Omaha, Nebraska

Adrienne Santiago
Career College of Northern Nevada
Sparks, Nevada

Greg Siepert
Hutchinson Community College
Hutchinson, Kansas

Ryan Woehl
Central Community College
Columbus, Nebraska

What's New?

This edition of *Math for Welders* contains numerous enhancements to help students succeed:

- Additional material on measurement and reading US customary and metric rulers has been added to Unit 15, Linear Measure, to ensure students have a strong foundation.

- Information on adding, subtracting, multiplying, and dividing feet and inches has also been added to Unit 15 to augment foundational coverage.

- Additional practice problems have been added to Unit 15 to reinforce these enhanced foundational topics.

- Drawings have been added in Unit 15 to illustrate the new material.

- Unit 22, Bending Metal, has been revised with improved explanations and examples.

- The problems in Unit 22 have been rearranged into a more logical sequence.

- New problems have been added to Unit 22.

- All the illustrations in *Math for Welders* have been redrawn for a more uniform, professional appearance.

- A second color has been used throughout the text to help clarify formulas and math examples, emphasize important content, and increase visual interest.

Brief Contents

Contents

SECTION 2 | COMMON FRACTIONS

SECTION 3 | DECIMAL NUMBERS

SECTION 1
Whole Numbers

After studying this section, you will be able to:

- Practice good math work habits.
- Explain the Arabic number system.
- Round off whole numbers.
- Define what is meant by a denominate number.
- Perform addition of whole numbers.
- Perform subtraction of whole numbers.
- Perform multiplication of whole numbers.
- Perform division of whole numbers.
- Check answers for four operations.

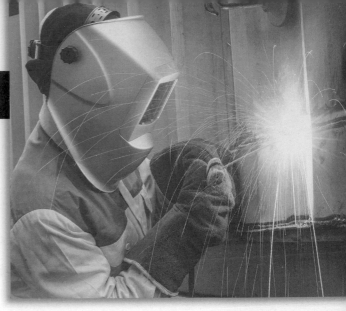

UNIT 1

Introduction to Whole Numbers

Key Terms

approximate numbers

Arabic number system

decimal number system

denominate numbers

place value

rounding

Introduction

A mastery of mathematics is one of the skills expected of you as a welder. Fortunately, the method of acquiring math skills is no different than any other skill. You learn the principles and then practice them in repeated applications until they become second nature.

What about the questions "Do I need to know math if I know how to use a calculator?" and "Is it necessary to develop a skill in mental calculation and a mastery of math facts?" There are practical reasons why the answer to these questions is *yes*. First, your employer and supervisor will expect you to be knowledgeable in math. They will expect that, as a tradesperson, you will bring basic math skills and knowledge of math facts to the job. If you begin your training in math by relying on a calculator, you will do yourself a great disservice. A calculator does not give you the confidence and understanding that comes with building a skill.

Also, when a group of welders is discussing the math aspects of a job (dimensions, weights, volumes, costs, etc.), you must be able to understand and contribute to the discussion. If your math skills are weak, you will likely feel uncomfortable about joining in on this part of your work. Besides, there is certain admiration granted to a tradesperson by fellow workers who recognize that the person is clearly proficient in math.

Math Work Habits

Although math is almost entirely a mental activity, there is a small but important tangible component to it. All your calculations, dimensions, and notes must be readable. Correct results are, of course, the final product of math, but the calculations should be done with a high degree of neatness and care. Orderly work in math is an important factor in arriving at correct answers. The following are some useful suggestions:

- Write your numbers fairly large.
- Write the numbers 4 and 9 carefully. When poorly written, these numbers are easily confused.

- Try writing 7 as 7̸. The number 7, when handwritten, can sometimes be confused with 1 or 2. The slash helps to avoid this confusion.
- Always carry a thick, sharp pencil while in the shop or in the field.
- When dealing with numbers, do not rush. In fact, deliberately slow yourself down. If you consider the amount of time it takes to turn out a job in the shop, there is no measurable gain in time by calculating and writing figures quickly.

Our Number System

Our number system is sometimes called the *Arabic number system*, because this system was developed in Arab culture. It is a system based on ten digits—0, 1, 2, 3, 4, 5, 6, 7, 8, 9. Our number system is also often referred to as the *decimal number system*. The word *decimal* is a word derived from Latin meaning *based on ten*.

Practically any number can be expressed by arranging these numbers in a certain order. When a series of digits is written, such as 2,497, each of the numbers in that group takes on a certain value based on its *place value* in the lineup. In the example of 2,497, the 7 is considered a 7 because it is in the ones (or units) position. The 9 has a value of 90 because it is in the tens position. You can think of it as $9 \times 10 = 90$. The 4, because of its place in the lineup, has a value of 400. It is in the hundreds column, so think of it as $4 \times 100 = 400$. The 2 is placed in the thousands position and has a value of 2,000.

Goodheart-Willcox Publisher

Each of the positions in a line of figures has a value and a name. Listed below are some of the names and their place position.

To help make numbers easier to read, a comma is usually placed after every

Goodheart-Willcox Publisher

third digit counting from the right. Here is an example. The number 3,894,076,215 is divided by commas and is read as three billion, eight hundred ninety-four million, seventy-six thousand, two hundred fifteen. The 3 has a place value of three billion. The 8 has a place value of eight hundred million, the 9 has a place value of ninety million, and so on.

Rounding Numbers

There are some occasions in math where extreme accuracy is not required. Cost estimates for most jobs are not usually exact. Precise weights of large weldments are often not necessary. In these and many other situations, figures may be rounded. ***Rounding*** is an approximating or adjusting of a number by increasing or decreasing a significant digit and cutting off or reducing to zero the least significant digit or digits. Since rounded figures are not perfectly accurate, they are called ***approximate numbers***.

How to Round Numbers

Here is a method for rounding numbers:

1. First, determine to what place value you are rounding. For example, you may be asked to round to the nearest hundred or thousand.
2. Place a little tick mark over the digit in that place position.
3. If the figure to the right of that number is five or more, increase the ticked number by one.
4. If the figure to the right is less than five, do not change the ticked number.
5. Lastly, all the numbers to the right of the rounded number are then replaced with zeros.

For example, a quiz question says to round 17,289,364 to the nearest ten thousand. After identifying the number in the ten thousand place, draw a tick mark over it.

$$17,2\overset{\checkmark}{8}9,364$$

Since the digit to the right of the 8 in the ten thousand place is 5 or more, the 8 is rounded up to 9.

$$17,2\overset{9}{8}9,364$$

The final step is to replace the numbers to the right of the rounded number with zeros.

$$17,29\overset{0000}{\cancel{9,364}}$$

Therefore, rounding 17,289,364 to the nearest ten thousand results in 17,290,000.

Denominate Numbers

Despite its long name, denominate numbers are simple. A ***denominate number*** is a number with a unit of measurement attached. For example, *4 feet* is a denominate number. The number *4*, by itself, is not a denominate number. Once you add a description or unit to a number showing that it represents some kind of measurement, it is classified as a denominate number.

The descriptive word *denominate* comes from the root word *denomination*. In this case, denomination means a group of units of measurement. Other examples of denominate numbers are 6 gallons, 5 3/16″, 186 lb, 1800°F, and 55 miles per hour. Each number is attached to a unit of measurement, such as volume, length, mass, temperature, or speed.

Since welders work almost entirely with denominate numbers, knowing how to work with these numbers is important. There are certain rules to follow. For now, remember that when an addition or subtraction question is expressed in denominate numbers, your answer must also be a denominate number.

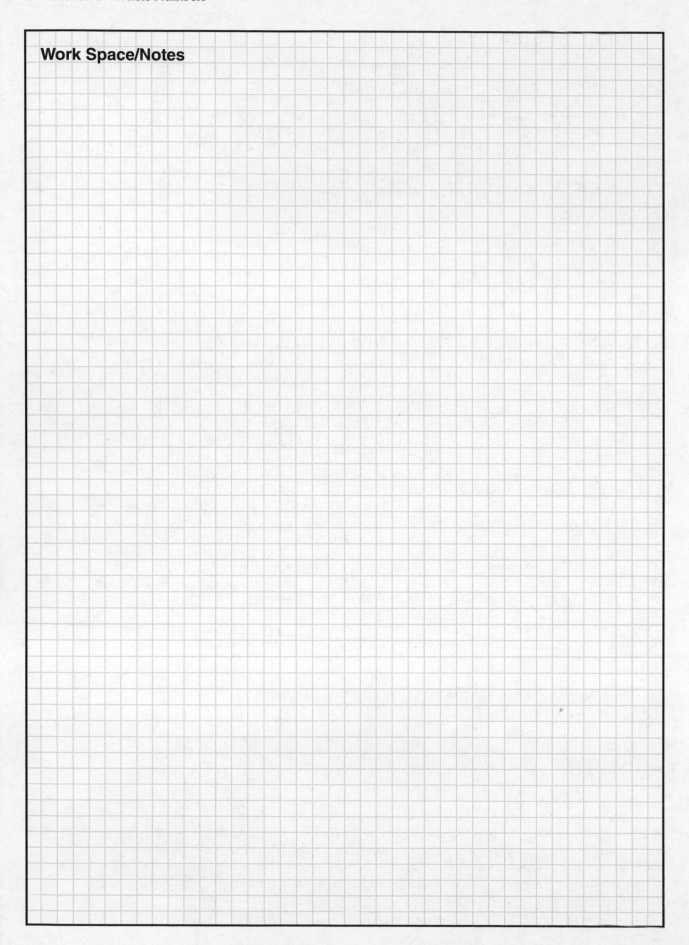

Work Space/Notes

Name _____ **Date** _____ **Class** _____

Label the place value of each digit in the following numbers.

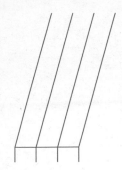

1. 471

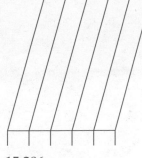

2. 15,286

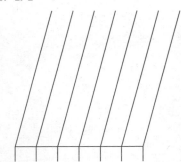

3. 349,015

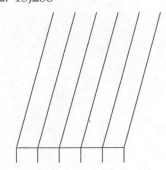

4. 27,636

Round the following numbers to the nearest hundred.

5. 694,701

6. 849

7. 17,213

8. 9,999

9. 29,897

10. 4,509

11. 945,999

12. 6,482,519

Round the following numbers to the nearest thousand.

13. 11,195

14. 449,561

15. 449,156

16. 85,028

17. 1,294,637

18. 9,195

19. 13,000,900

20. 10,001

21. In the following list, circle each denominate number.

250 psi	$7.65
11,500 ft^2	$7.65 per hour
five acres	40" per minute
17.5 tons	12
1,895	12 dozen
0.525	

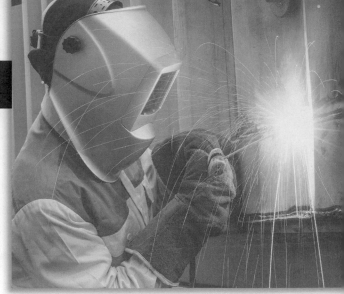

UNIT 2

Addition of Whole Numbers

Key Terms

addition

basic operations

carrying

sum

Introduction

Whole numbers are manipulated in arithmetic by four *basic operations*—addition, subtraction, multiplication, and division. The most widely used of these four operations is addition. You will learn about addition in this unit. The other three basic operations will be studied in following units.

Method Used to Add Whole Numbers

Addition is the process of combining two or more individual numbers to form a single number that usually has a higher value than any of the previously individual numbers. Whole numbers are added by stacking the numbers to be added in a column with all the units on the right side lined up one on top of the other.

<pre>
        units   units   units
          ↓   ↓   ↓
        4,956   27   354
           21   192   2,032
          191   11   111
          256   1,542   9,461
      + 10,431   + 5   + 17
      ─────────   ───────   ───────
</pre>

It is important that all of the digits for each place value line up exactly above other digits of the same place value. Refer to the illustration of place values in Unit 1. Once each column has been calculated, a final answer is reached. The final answer in addition equations is called the *sum*.

Carrying

Only one digit can occupy a single column. What happens when the sum of numbers in a single column reaches or exceeds 10 and has a value that occupies two columns? In the following example, adding the numbers in the units column results in 15. The 5 is placed under the units column and the 1 is carried to the next column. Moving the second digit of a sum up a column in value is called *carrying*. Whenever a single column's sum totals a value of 10 or more, move the higher valued numbers one column to the left.

$$
\begin{array}{r}
^{1\,2\,2\,1} \\
9{,}966 \\
21 \\
991 \\
256 \\
+\ 94{,}531 \\
\hline
105{,}765
\end{array}
$$

When you carry a number, write the carried number at the top of the next column and make it smaller than your other digits. Next, add the tens column. The result is 26. Write in the number 6, and carry the 2 to the hundreds column. Continue in this manner until all columns are added. Always check your accuracy.

Look for Tens

When adding a column, search for combinations of numbers that add up to 10. In the single-column example shown, 8 plus 2 equals 10, and 1 plus 9 equals 10 (those add up to 20, so far), and 4 plus 6 equals 10 (totaling 30). The remaining digit is 7, so the total is 37. You will find it easier doing the mental calculations this way. Be sure to write tick marks next to the numbers as you add them or you may forget which ones you have already added:

$$
\begin{array}{r}
9 \\
7 \\
2\ \checkmark \\
6 \\
4 \\
1 \\
+\ 8\ \checkmark \\
\hline
37
\end{array}
$$

Checking Addition by Adding Up and Down

Always check your math work. If you added by going up the column, then check it by adding going down.

Mark the Answer

When you arrive at a final answer, mark it in a very clear manner. Try boxing it in, like this $\boxed{248}$, or double underlining it 248. This is especially important if you have several numbers on a sheet mixed in with previous, unrelated calculations. This is sometimes the situation when doing calculations in the shop or in the field.

Denominate Numbers

Just as important as marking the right number is including the right unit in the answer. When marking answers, be aware of denominate numbers. If a question is expressed in inches, your answer should also be in inches.

When adding denominate numbers, be certain that all the numbers have the same unit. Only denominate numbers having the same type of unit can be added. If an addition equation contains denominate numbers with dissimilar units, decide which unit is the most convenient for addition. Convert the numbers to the convenient unit. Then add the numbers.

Compatible Units	Incompatible Units
4″	9 cm
+ 7″	+ 4″
11″	

The equation with 9 cm and 4″ cannot be calculated as it is. One of these denominate numbers must be converted to the other denominate number's unit of measurement. Then this equation can be completed. Conversion between different units will be covered later in the text.

This concept of sharing common units is also applicable to subtraction equations. Denominate numbers must have the same units in order to subtract from one another. To refresh the concept of denominate numbers, review the Denominate Numbers section in Unit 1.

Work Space/Notes

Name _____ **Date** _____ **Class** _____

Add the following number groupings. Show all your work. Be certain the columns line up. Box your answers.

1.	235	2.	6,241	3.	964
	471		7,356		33,508

4.	102	5.	75	6.	11,009
	268		5,491		7,489,621
	754		67,811		28,912,305
	888		604		47,951
	937		15,473		1,997

7.	8	8.	46	9.	167,540
	9		365		22,814
	7		3		499,863
	6		1,588		200
	2		331		98,157
	4		91		5
	1		942		37,254

Add the following number groupings. Box your answers.

10. 7 + 9 + 1 + 3 + 3 + 8 + 5 =

11. 904 + 214 + 22 =

12. 18,961 + 718 + 6,800 =

13. 6,525 + 7,182 + 293 =

14. 966 + 372 + 165 + 638 + 300 + 200 =

15. 731 + 82 + 234 + 2,699 + 523 + 64 =

16. 18 + 444 + 27,981 + 1,234 + 75,211 + 7 =

17. 21,987 + 101,001 + 622 + 9 + 36,299 + 981,202 =

18. 4 + 144 + 414,411 + 69,835 + 641,538 + 99 =

19. Three pieces of double extra-strong pipe in inventory have the following lengths: 27", 42", and 19". What is the total length of double extra-strong pipe in inventory?

20. Three welded frames are shipped by truck to a customer. The smallest frame weighs 985 lb, the next frame weighs 2,891 lb, and the largest weighs 3,257 lb. What is the total weight?

21. Tregaskiss, Inc., a manufacturer of welding rods, produced the following quantities: 9,622 for January, 72,450 for February, and 284,361 for March. What was the total production for the three months?

22. According to statistics, a coal mining company's production for each of the past six years has been 9,253,589 tons; 9,405,740 tons; 8,713,903 tons; 8,001,878 tons; 7,464,890 tons; and 7,791,261 tons. How many tons of coal did the company produce in the past six years?

23. Alexander Lincoln worked the number of hours shown below for the month of December. What is the total number of hours he worked for the month?

December								
Week	Date	Hours	Week	Date	Hours	Week	Date	Hours
1	1	8	2	10	8	4	21	7
1	2	8	2	11	6	4	22	8
1	3	8				4	23	6
1	4	8	3	14	4	4	24	3
1	5	4	3	15	8			
			3	16	8	5	28	5
2	7	6	3	17	8	5	29	6
2	8	8	3	18	7	5	30	9
2	9	8	3	19	5	5	31	3

Goodheart-Willcox Publisher

Name _____   **Date** _____   **Class** _____

Use the information below to answer the questions that follow. A manufacturing company with several departments is compiling information for its Human Resources records. The current list is as follows:

2,816 men and 1,042 women in the Fabricating Department

110 men and 119 women in the Repair and Maintenance Department

39 men and 63 women in the Administration Department

367 men and 457 women in the Engineering Department

98 men and 85 women in the Sales and Service Department

24. How many men are employed by the company?

25. How many women are employed by the company?

26. What is the total number of people employed?

27. What is the total length of angle iron in the following illustration?

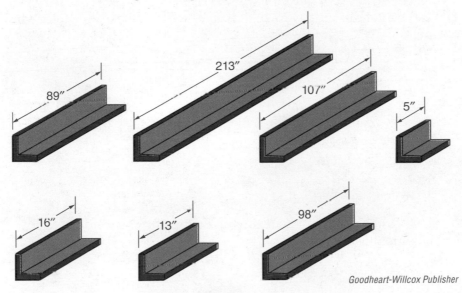

Goodheart-Willcox Publisher

28. What is the total length of the notched plate shown in the following illustration?

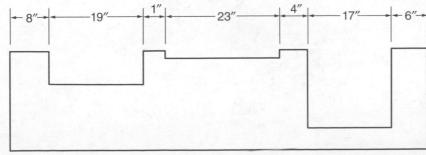

Goodheart-Willcox Publisher

29. What is the total outside length of the angle iron frame shown in the following illustration?

30. What is the total inside length of this frame?

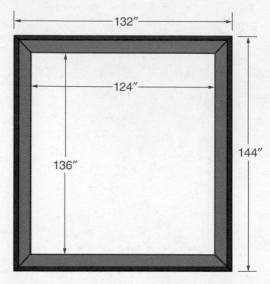

Goodheart-Willcox Publisher

31. What is the total distance between the centers of the holes in the following illustration?

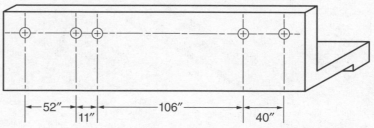

Goodheart-Willcox Publisher

32. What is the total length of square tubing required for the weldment shown in the following illustration?

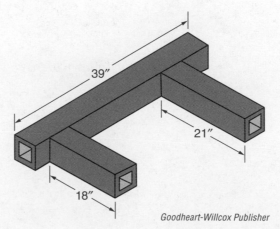

Goodheart-Willcox Publisher

Name _____   **Date** _____   **Class** _____

33. This pie chart illustrates the number of hours required to complete various phases of Job Order #8805. What is the total number of hours required to complete the job?

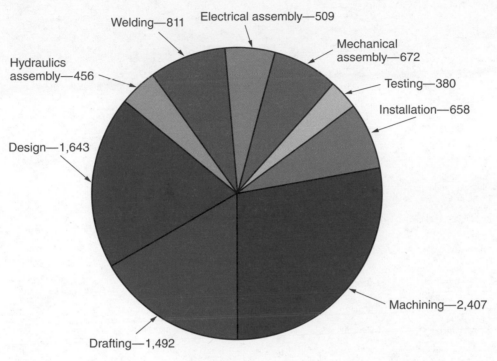

Welding—811   Electrical assembly—509

Mechanical assembly—672

Hydraulics assembly—456

Testing—380

Installation—658

Design—1,643

Machining—2,407

Drafting—1,492

34. How many tons of steel are produced in total by the countries listed in the following table?

Annual Steel Production	
Country	**Production in Tons**
US	67,746,000
Canada	14,825,000
China	35,620,000
Slovakia	15,319,000
France	21,346,000
Germany	41,662,000
Italy	25,085,000
Japan	101,708,000
Poland	15,112,000
Romania	13,193,000
Russia	148,980,000
UK	15,667,000

Work Space/Notes

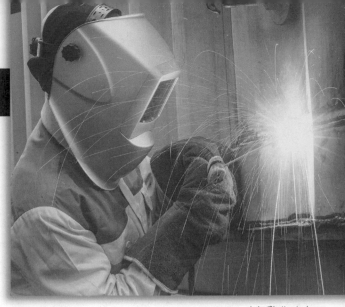
anekoho/Shutterstock.com

Subtraction of Whole Numbers

Key Terms

borrow

difference

minuend

remainder

subtraction

subtrahend

Introduction

Subtraction is one of the four basic math operations. *Subtraction* is the process of finding the difference between two numbers. As with most math operations, there are new terms to learn. Below are the names of the parts of a subtraction equation:

$$1,347 \longleftarrow \textbf{Minuend}$$

$$\underline{-\ 26} \longleftarrow \textbf{Subtrahend}$$

$$1,321 \longleftarrow \textbf{Difference}\ \text{or}\ \textbf{Remainder}$$

The larger number in a subtraction equation is the *minuend*. In this equation, the minuend is the top number, and the subtrahend is below it. The *subtrahend* is the number that will be removed from the minuend. The resulting number is the *difference*, which may also be called the *remainder*.

Method Used to Subtract Whole Numbers

Two whole numbers are subtracted by stacking them one on the other with the units column lined up on the right side. The larger number belongs on the top position (minuend). Begin subtracting at the units column and work your way through the problem, column by column. Here are two examples.

$$\begin{array}{r} 84 \\ -\ 23 \\ \hline 1 \end{array} \qquad \begin{array}{r} 5,794 \\ -\ 62 \\ \hline 2 \end{array}$$

It is important to begin subtraction in the units column. Numbers in the lower columns can affect the values of the numbers in the higher columns. In the preceding examples, the subtraction in the units columns has been completed successfully, allowing subtraction to continue in the higher columns.

$$
\begin{array}{r}
84 \\
-23 \\
\hline
61
\end{array}
\qquad
\begin{array}{r}
5{,}794 \\
-62 \\
\hline
5{,}732
\end{array}
$$

Borrowing

As previously mentioned, calculating subtraction equations must begin at the units column. This is because lower value columns can affect higher value columns. You will often encounter a situation like this:

$$
\begin{array}{r}
47 \\
-19
\end{array}
$$

The units column displays 7 minus 9. Since 9 cannot be subtracted from 7, you must **borrow** a number from the place value to the left. To borrow a number, subtract 1 from the number in the column to the left and add that 1 to the column that needs it. In this example, subtract 1 from the 4 in the tens column. Draw a line through the 4 and replace it with 3. Now, place a tiny 1 next to the 7 to create the number 17.

$$
\begin{array}{r}
\overset{3}{\cancel{4}}\overset{1}{7} \\
-19
\end{array}
$$

Now, subtract 9 from 17 to get 8. Then, to finish the problem, subtract 1 from 3 to get 2.

$$
\begin{array}{r}
\overset{3}{\cancel{4}}\overset{1}{7} \\
-19 \\
\hline
28
\end{array}
$$

Here is a more complicated example:

$$
\begin{array}{r}
1{,}007 \\
-99
\end{array}
$$

In the units column, you cannot subtract 9 from 7. As in the previous example, you look to the tens column to borrow a 1 and find there is no number to borrow. You then have to go to the hundreds column and still there is no number to borrow. Finally, in the thousands column, you can borrow a 1. By borrowing 1, the existing 1 becomes 0. Draw a line through the existing 1, and it becomes 0. Now, place a tiny 1 next to the 0 in the hundreds column to create the number 10.

$$
\begin{array}{r}
\overset{0}{\cancel{1}}\overset{1}{,}007 \\
-99
\end{array}
$$

Now there is a 10 in the hundreds column. Borrow 1 from that 10 and reduce it to 9 by crossing out the 10 and replacing it with a 9. Add the borrowed 1 to the 0 in the tens column.

$$
\begin{array}{r}
9 \\
0\ 1\ 1 \\
\cancel{1,0}07 \\
-\ 99 \\
\end{array}
$$

Now, there is a 10 in the tens column. Borrow 1 from the 10 in the tens columns and reduce it to 9. Add 1 to the 7 in the units column, thereby changing it to 17. You can now begin the problem by subtracting 9 from 17 in the units column.

$$
\begin{array}{r}
9\ 9 \\
0\ 1\ 1\ 1 \\
\cancel{1,0}07 \\
-\ 99 \\
\hline
908 \\
\end{array}
$$

Checking Subtraction by Adding

Always check your accuracy. Check your subtraction work by adding the difference and the subtrahend. The total should equal the minuend.

Minuend	198,462
Subtrahend	− 76,291
Difference	122,171
Subtrahend	+ 76,291
Minuend	198,462

Denominate Numbers

Concerning denominate numbers, subtraction equations follow the same rule as addition equations. Denominate numbers in a subtraction equation must each have the same unit of measurement. If two denominate numbers with different units are given in a subtraction equation, convert one of the units to the other.

Subtraction Terminology

When speaking or writing about subtraction, there are a number of ways to communicate the situation. If you are not familiar with these expressions, it could be difficult to figure out just what is being subtracted from what. The following are ways to state a subtraction equation:

    24 take away 11

    24 less 11

    24 minus 11

    24 subtract 11

    24 reduced by 11

    11 from 24

    Find the difference between 24 and 11

All of these expressions are ways of writing the equation 24 − 11.

Work Space/Notes

Name _____ **Date** _____ **Class** _____

Subtract the following number groupings. Show all your work. Be certain the columns line up.
Box your answers.

1.   875
    − 211

2.   1,280
    − 1,120

3.   9,706
    − 7,960

4.   74,240
    − 659

5.   21,008
    − 989

6.   12,345
    − 6,789

7.   45,000
    − 2,370

8.   894,216
    − 735,094

9.   1,019,471
    − 4,048

10. 847 − 305 =

11. 6,544 minus 822 =

12. Take 28,001 away from 29,477 =

13. Find the difference between 18,234 and 11,012 =

14. Find the result when 707 is taken away from 747 =

15. What is the difference between 74 and 47 =

16. Subtract 7,641 from 22,350 =

17. 184,555 less 90,915 =

18. Reduce 19,486 by 215 =

19. Advance Metal Works had 1,717 brackets in stock on February 15th. One month later, there were 912 remaining in inventory. How many brackets were used during the month?

Name _____   **Date** _____   **Class** _____

Use this information to answer the questions that follow. A fabricating shop had the following steel in inventory:

15,285 lb of 22 gage steel

37,549 lb of 16 gage steel

89,041 lb of 14 gage steel

One large customer order used the following quantities:

6,102 lb of 14 gage steel

2,819 lb of 16 gage steel

11,910 lb of 22 gage steel

20. What weight of 22 gage steel remained after the order was completed?

21. What weight of 16 gage steel remained after the order was completed?

22. What weight of 14 gage steel remained after the order was completed?

23. What was the total weight of the customer order?

Use this information to answer the questions that follow. Proto Mfg. produced the following welding tips in one week:

Tip Size	Quantity Produced
#68	118,295
#51	7,050
#35	892

Goodheart-Willcox Publisher

The following quantities were shipped to various welding supply houses:

Tip Size	Quantity Shipped
#35	0
#68	73,460
#51	5,070

Goodheart-Willcox Publisher

24. How many #68 tips remain at Proto?

25. How many #51 tips remain at Proto?

26. How many #35 tips remain at Proto?

27. The original design for a large machine calls for a frame of structural steel tubing that weighs 4,715 lb. An engineering change was later approved, adding additional stiffeners. The final weight of the frame was 5,128 lb. What is the weight of the additional stiffeners?

Name _____   **Date** _____   **Class** _____

28. A survey determined that there were 261,298 service stations in the United States. After four years, a repeat of the survey showed this number had declined by 48,424. Among the stations still in business after four years, there were 54,631 that did not use welding equipment. How many service stations still used welding equipment at the time of the second survey?

29. A coal car was redesigned by replacing some of the steel with aluminum. The designers wanted to reduce its weight and also increase its payload to 110 tons. The total weight of the redesigned car was 42,695 lb. The amount of aluminum used was 9,860 lb. The remainder of the redesigned car was steel. What weight of steel was used?

30. Three pieces are cut from a length of cold-rolled steel. What length remains? Ignore loss of material due to cutting.

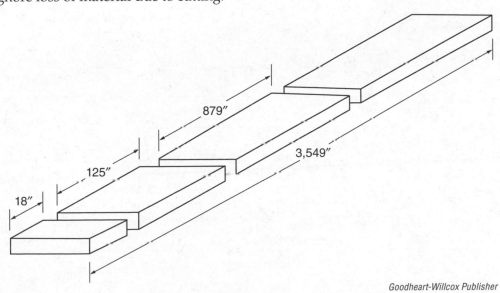

Goodheart-Willcox Publisher

31. What is the distance between the centers of the holes in the piece shown in the following illustration?

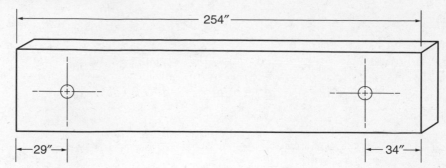

Goodheart-Willcox Publisher

32. What is the height of the vertical piece Ⓐ?

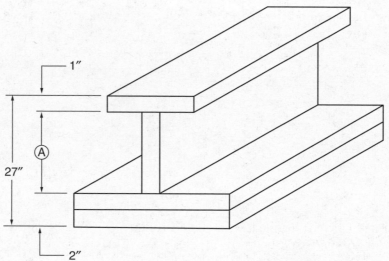

Goodheart-Willcox Publisher

33. What is the length of distance Ⓐ?

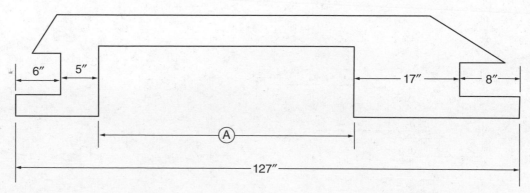

Goodheart-Willcox Publisher

Name _____   **Date** _____   **Class** _____

34. What is the difference in height (as indicated by Ⓐ) between the two pipes?

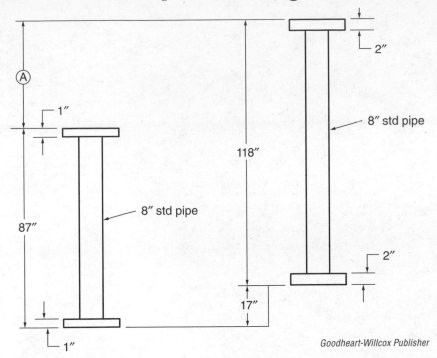

Goodheart-Willcox Publisher

Review the diagram and the supply room's pre-job inventory list.

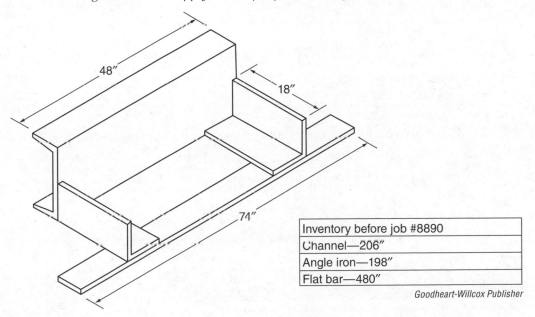

Inventory before job #8890
Channel—206"
Angle iron—198"
Flat bar—480"

Goodheart-Willcox Publisher

35. After this project is finished, how many inches of channel remain in the supply room?

36. After this project is finished, how many inches of angle iron remain in the supply room?

37. After this project is finished, how many inches of flat bar remain in the supply room?

Multiplication of Whole Numbers

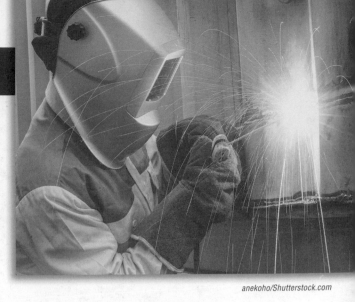

anekoho/Shutterstock.com

Key Terms

carry

multiplicand

multiplication

multiplier

product

Introduction

Multiplication can be thought of as a fast way to do addition. The process involves the repeated addition of a single number. This may seem odd at first, but look at it this way. If you were to add 3 + 3 + 3 + 3 + 3, the total, of course, would be 15. However, you could think of it as, "3 taken 5 times" or "5 times the number 3." This way of thinking works well for mental calculations of small numbers. It speeds up the process and simplifies the problem. Many people have memorized the multiplication facts for numbers 1 through 12 and these are often seen in chart form as multiplication tables.

Multiplication Table

2 × 1 = 2	3 × 1 = 3	4 × 1 = 4	5 × 1 = 5	6 × 1 = 6	7 × 1 = 7
2 × 2 = 4	3 × 2 = 6	4 × 2 = 8	5 × 2 = 10	6 × 2 = 12	7 × 2 = 14
2 × 3 = 6	3 × 3 = 9	4 × 3 = 12	5 × 3 = 15	6 × 3 = 18	7 × 3 = 21
2 × 4 = 8	3 × 4 = 12	4 × 4 = 16	5 × 4 = 20	6 × 4 = 24	7 × 4 = 28
2 × 5 = 10	3 × 5 = 15	4 × 5 = 20	5 × 5 = 25	6 × 5 = 30	7 × 5 = 35
2 × 6 = 12	3 × 6 = 18	4 × 6 = 24	5 × 6 = 30	6 × 6 = 36	7 × 6 = 42
2 × 7 = 14	3 × 7 = 21	4 × 7 = 28	5 × 7 = 35	6 × 7 = 42	7 × 7 = 49
2 × 8 = 16	3 × 8 = 24	4 × 8 = 32	5 × 8 = 40	6 × 8 = 48	7 × 8 = 56
2 × 9 = 18	3 × 9 = 27	4 × 9 = 36	5 × 9 = 45	6 × 9 = 54	7 × 9 = 63
2 × 10 = 20	3 × 10 = 30	4 × 10 = 40	5 × 10 = 50	6 × 10 = 60	7 × 10 = 70
2 × 11 = 22	3 × 11 = 33	4 × 11 = 44	5 × 11 = 55	6 × 11 = 66	7 × 11 = 77
2 × 12 = 24	3 × 12 = 36	4 × 12 = 48	5 × 12 = 60	6 × 12 = 72	7 × 12 = 84

Goodheart-Willcox Publisher

(continued)

$8 \times 1 = 8$	$9 \times 1 = 9$	$10 \times 1 = 10$	$11 \times 1 = 11$	$12 \times 1 = 12$
$8 \times 2 = 16$	$9 \times 2 = 18$	$10 \times 2 = 20$	$11 \times 2 = 22$	$12 \times 2 = 24$
$8 \times 3 = 24$	$9 \times 3 = 27$	$10 \times 3 = 30$	$11 \times 3 = 33$	$12 \times 3 = 36$
$8 \times 4 = 32$	$9 \times 4 = 36$	$10 \times 4 = 40$	$11 \times 4 = 44$	$12 \times 4 = 48$
$8 \times 5 = 40$	$9 \times 5 = 45$	$10 \times 5 = 50$	$11 \times 5 = 55$	$12 \times 5 = 60$
$8 \times 6 = 48$	$9 \times 6 = 54$	$10 \times 6 = 60$	$11 \times 6 = 66$	$12 \times 6 = 72$
$8 \times 7 = 56$	$9 \times 7 = 63$	$10 \times 7 = 70$	$11 \times 7 = 77$	$12 \times 7 = 84$
$8 \times 8 = 64$	$9 \times 8 = 72$	$10 \times 8 = 80$	$11 \times 8 = 88$	$12 \times 8 = 96$
$8 \times 9 = 72$	$9 \times 9 = 81$	$10 \times 9 = 90$	$11 \times 9 = 99$	$12 \times 9 = 108$
$8 \times 10 = 80$	$9 \times 10 = 90$	$10 \times 10 = 100$	$11 \times 10 = 110$	$12 \times 10 = 120$
$8 \times 11 = 88$	$9 \times 11 = 99$	$10 \times 11 = 110$	$11 \times 11 = 121$	$12 \times 11 = 132$
$8 \times 12 = 96$	$9 \times 12 = 108$	$10 \times 12 = 120$	$11 \times 12 = 132$	$12 \times 12 = 144$

Goodheart-Willcox Publisher

1	2	3	4	5	6	7	8	9	10	11	12
2	4	6	8	10	12	14	16	18	20	22	24
3	6	9	12	15	18	21	24	27	30	33	36
4	8	12	16	20	24	28	32	36	40	44	48
5	10	15	20	25	30	35	40	45	50	55	60
6	12	18	24	30	36	42	48	54	60	66	72
7	14	21	28	35	42	49	56	63	70	77	84
8	16	24	32	40	48	56	64	72	80	88	96
9	18	27	36	45	54	63	72	81	90	99	108
10	20	30	40	50	60	70	80	90	100	110	120
11	22	33	44	55	66	77	88	99	110	121	132
12	24	36	48	60	72	84	96	108	120	132	144

Goodheart-Willcox Publisher

To use this chart, the product of multiplying each number along the top row by each number in the left column is found where the row and column intersect. For example, to find the answer to 8 times 4, locate their intersecting squares. Following column 8 down to row 4 shows the answer to be 32. Likewise, following column 4 down to row 8 also results in 32.

Incidentally, there is no special reason why the multiplication table stopped at 12. It is probably because most people did not want to memorize beyond that point. Even in this age of electronic calculators, you should commit the multiplication table to memory.

Method Used to Multiply Whole Numbers

To begin, you should know the appropriate terminology. The upper number in the multiplication equation is the *multiplicand*. This is the number that is being added repeatedly. The number below the multiplicand is the *multiplier*. This number represents how many times the multiplicand is repeatedly added. The result of a multiplication equation is the *product*.

$$132 \longleftarrow \textbf{Multiplicand}$$
$$\underline{\times 23} \longleftarrow \textbf{Multiplier}$$
$$3,036 \longleftarrow \textbf{Product}$$

Always line up the figures on the right side in the units column. Multiply every digit in the multiplicand by the number in the units column of the multiplier. Start multiplying the multiplicand at the units column. In the following example, it starts with the 3 in the multiplier's units column and the 2 in the multiplicand's units column. Refer to the multiplication tables if you need to refresh your skills as you study the following examples.

$$132$$
$$\underline{\times 23}$$
$$396$$

Next, multiply every digit in the multiplicand by the 2 in the multiplier's tens column. Place the results of this multiplication below the first results. Indent the second results so they line up directly under the 2 in the multiplier. Then, add the results of the two multiplication operations.

$$132$$
$$\underline{\times 23}$$
$$396$$
$$\underline{264\ \ }$$
$$3,036$$

In multiplication problems, you often have to *carry* numbers. This was also done in addition equations when the sum of a single column was 10 or higher. Since only one digit can occupy a column, a number was carried to the next column. See the example below.

$$298$$
$$\underline{\times 27}$$

In the first operation, $7 \times 8 = 56$. Place the 6 in the answer and write a small 5 directly above the 9.

$$\overset{5}{298}$$
$$\underline{\times 27}$$
$$6$$

Now, multiply 7×9 to get 63 and add the carried 5 to it, arriving at 68. Place 8 in the answer and a small 6 above the 2.

$$\overset{65}{298}$$
$$\underline{\times 27}$$
$$86$$

Multiply 7×2 to get 14. Add the carried 6 to the 14 to get 20. Write 20 in the answer. Always check your accuracy.

$$
\begin{array}{r}
{}^{6\,5} \\
298 \\
\times 27 \\
\hline
2086
\end{array}
$$

For larger equations like this one, neatness is important. When multiplying the next set of numbers, write the carried numbers above the previously used carried numbers. Keep careful track of these numbers, so you do not add the wrong numbers. In larger equations, you may want to cross out carried numbers after they have been used in a calculation.

$$
\begin{array}{r}
{}^{1\,1} \\
{}^{6\,5} \\
298 \\
\times 27 \\
\hline
2086 \\
596 \\
\hline
8{,}046
\end{array}
$$

Accurate Alignment

Be sure to line up the numbers accurately in your answer. You may include right-hand zeros or X's to help keep the figures aligned. These zeros and X's occupy the space of indented columns.

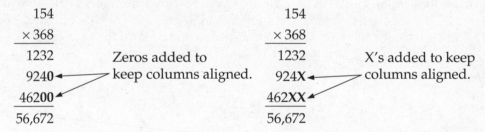

Notice the bold zeros and X's in the examples above. Including such characters will help to keep your columns properly aligned. Leaving those spaces open may lead to column confusion and incorrect results.

Checking Multiplication by Reversing Positions

A method for checking your multiplication work is to reverse the positions of the multiplier and multiplicand and repeat the problem. Performing the equation each way should result in the same product. The numbers obtained by multiplying each digit of the multiplier with the multiplicand will be different than the numbers from the first equations. However, these different numbers should still add up to the same final product as the original equation.

Original Equation   **Reverse Position Equation**

643   972
× 972   × 643

1286 ◄——— Different ———► 2916
45010 ◄——— Different ———► 38880
578700 ◄—— Different ———► 583200
624,996 ◄——— Same ———► 624,996

As shown in the example, the initial multiplication results may be different, but the final product should match the original answer.

Smaller Multiplier

It does not matter which of the two numbers is positioned as the multiplier or multiplicand. Both will provide the same results. However, it may be easier to place the smaller number as the multiplier. This results in fewer rows of numbers to add.

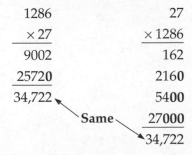

1286   27
× 27   × 1286
9002   162
25720   2160
34,722   5400
      Same   27000
              34,722

Denominate Numbers

How denominate numbers work in an equation is based on the operation of the equation. For instance, we already know that denominate numbers must have the same units to be added or subtracted. However, in multiplication and division equations, denominate numbers do not need to have the same units. In a multiplication equation, the units of measurements are combined.

When denominate numbers with different units of measurement are multiplied, they form a new unit of measurement. See the following example.

$$12 \text{ lb} \times 7' = 84 \text{ ft-lb}$$

While the numbers are multiplied, the pound and foot units are combined to form foot-pounds (ft-lb). These compound units can be formed by multiplication and undone by division. When denominate numbers sharing the same unit of measurement are multiplied, they form a squared unit. See the example below.

$$7'' \times 8'' = 56 \text{ in}^2$$

Multiplying two numbers with inch measurements forms a product measured in inches squared (in^2). More information on denominate numbers in multiplication equations will be covered in later units on measurement.

Name _____ **Date** _____ **Class** _____

Multiply the following number groupings. Show all your work. Be certain the columns line up. Box your answers.

1. $\begin{array}{r} 12 \\ \times\,34 \\ \hline \end{array}$

2. $\begin{array}{r} 98 \\ \times\,75 \\ \hline \end{array}$

3. $\begin{array}{r} 56 \\ \times\,70 \\ \hline \end{array}$

4. $\begin{array}{r} 659 \\ \times\,423 \\ \hline \end{array}$

5. $\begin{array}{r} 207 \\ \times\,845 \\ \hline \end{array}$

6. $\begin{array}{r} 913 \\ \times\,600 \\ \hline \end{array}$

7. $\begin{array}{r} 84{,}796 \\ \times\,56 \\ \hline \end{array}$

8. $\begin{array}{r} 93{,}253 \\ \times\,24 \\ \hline \end{array}$

9. $\begin{array}{r} 53{,}602 \\ \times\,35 \\ \hline \end{array}$

10. $253 \times 6{,}944 =$

11. $804 \times 5{,}022 =$

12. $68 \times 12{,}673 =$

13. $69{,}671 \times 36 =$

14. $12 \times 24 \times 36 =$

15. $18 \times 13 \times 20 =$

16. $109 \times 14 \times 34 =$

17. $263 \times 101 \times 24 =$

18. $468 \times 219 \times 153 =$

19. Lee's Welding produced 193 base frames. Each frame required 98 spot welds. What is the total number of spot welds for the entire job?

20. Ramido's Rapid Delivery Service delivered four pallets of rivets to Weld-Can Mfg. Each pallet contained 27 cartons and each carton contained 24 boxes. Each box contained 68 rivets. How many rivets were delivered?

21. Thirty-nine rows of studs will be welded to a metal platform. Each row will be 5″ apart. There will be 27 studs in each row. What is the total number of studs required?

22. A GMAW welder traveling at 17″ per minute was in constant use for 17 1/2 hours each day for 24 days. How many inches of weld were deposited in that time?

23. The roof of an arena built for the University of Iowa requires 5,247 plates to secure the roof trusses to the support beams. These plates vary in size and number of bolts required. The plates were flame cut to shape, then subcontracted to a machine shop for drilling the bolt holes. According to the information listed, what is the total number of bolt holes that need to be drilled?

1,728 plates with 6 holes	2,235 plates with 14 holes
295 plates with 5 holes	15 plates with 23 holes
964 plates with 9 holes	10 plates with 10 holes

Name _____   **Date** _____   **Class** _____

Use the following information to answer questions 24 and 25. The supplies listed must be transported from the ground floor storage to the top floor. Workers plan to use a freight elevator with a capacity of 8,500 lb to send everything at once.

Steel Rods

Seven bundles of 17 rods each. Each rod is 4′ long and weighs 2 lb per foot.

Plates

Twelve stacks of 13 plates each. Each plate weighs 14 lb. The plates are stacked on four wooden pallets that weigh 39 lb each.

Angle Iron

Forty-two pieces. Each piece weighs 18 lb and is 4′ long.

Aluminum Tubing

Six bundles of 16 tubes. Each tube weighs 7 lb.

Four bundles of 12 tubes. Each tube weighs 9 lb.

Seven bundles of five tubes each. Each tube is 4′ long and weighs 2 lb per foot.

Four bundles of five tubes each. Each tube is 3′ long and weighs 2 lb per foot.

Pipe

Two bundles of 20 pipes each. Each pipe is 3′ long and weighs 2 lb per foot.

Seven bundles of 11 pipes each. Each pipe weighs 11 lb.

Hubs

Four cartons of 19 boxes each. Each box contains 6 hubs. The weight of each box including the six hubs is 13 lb. Each carton is on a wooden pallet that weighs 39 lb.

Operator

The elevator operator weighs 165 lb.

24. Will the workers be able to send the whole load in a single trip?

25. What is the total load?

Use this information to answer questions 26 through 29. A power line built for Louisiana Light and Power Co. had 619 towers constructed of various sizes of pipe. Specifications for the job are listed in the following chart.

Steel Pipe Requirements			
	Tower Legs	**Cross Arms**	**Stiffeners**
Material	ASTM A595	ASTM A595	ASTM A595
Size	2′ diameter	1½′ diameter	1′ diameter
Number per Tower	2	1	2
Length	93′ on 175 towers	102′	35′
	108′ on 300 towers		
	122′ on 144 towers		

Goodheart-Willcox Publisher

26. Calculate the total length of 2′ diameter pipe.

27. Calculate the total length of 1 1/2′ diameter pipe.

28. Calculate the total length of 1′ diameter pipe.

29. Find the total length of pipe for the entire project.

Name _____   **Date** _____   **Class** _____

Use the information shown in the following illustration to answer questions 30 and 31.
The illustration depicts equally sized rings that are welded together.

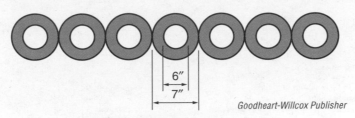

6"
7"

Goodheart-Willcox Publisher

30. What is the total length of the weldment?

31. What is the total weight of the weldment if each ring weighs 6 lb?

32. How many of the following GMAW welding tips can be produced in 67 working days at 9 hours per day?

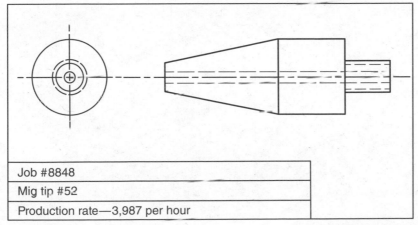

| Job #8848 |
| Mig tip #52 |
| Production rate—3,987 per hour |

Goodheart-Willcox Publisher

33. How many inches of chrome-plated tubing will be required to manufacture 289 tables?

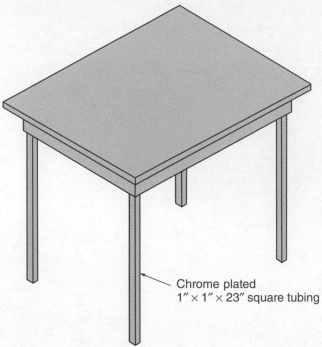

Chrome plated
1″ × 1″ × 23″ square tubing

Goodheart-Willcox Publisher

34. Each of the posts in this shaft support is welded all around. How many inches of weld will be applied to make 12,983 shaft supports?

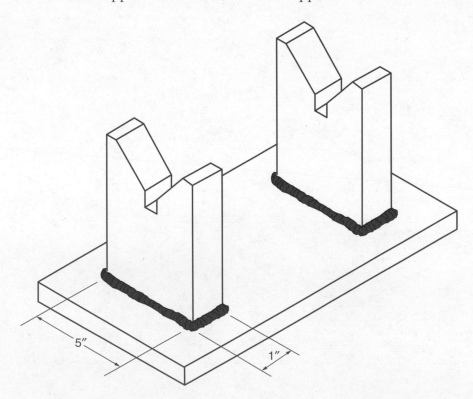

5″

1″

Goodheart-Willcox Publisher

Name _____   **Date** _____   **Class** _____

35. Review the following diagram. Calculate the weight of the weldment. Assume there is no filler added to the welds.

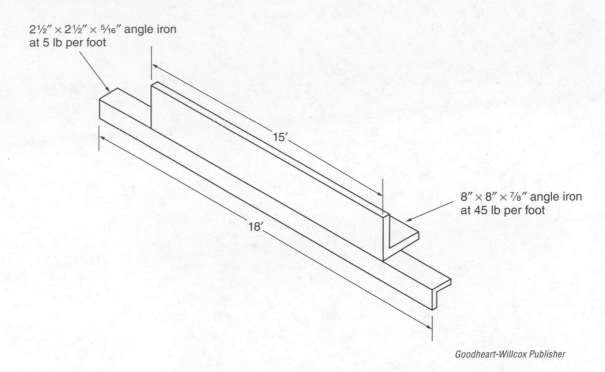

2½″ × 2½″ × ⁵⁄₁₆″ angle iron at 5 lb per foot

15′

18′

8″ × 8″ × ⅞″ angle iron at 45 lb per foot

Goodheart-Willcox Publisher

36. Review the following diagram. Calculate the weight of the weldment. Assume there is no filler added to the welds.

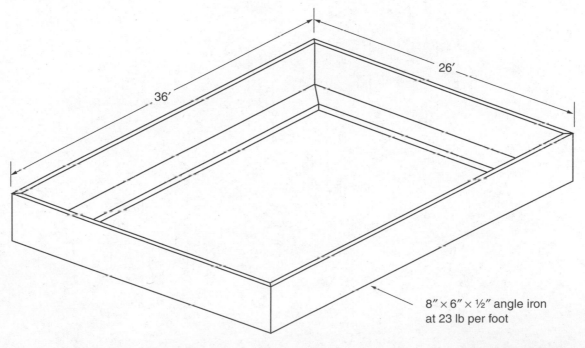

36′

26′

8″ × 6″ × ½″ angle iron at 23 lb per foot

Goodheart-Willcox Publisher

37. Review the following diagram. Calculate the weight of the weldment. Assume there is no filler added to the welds.

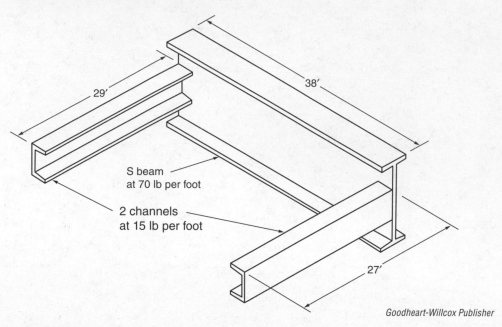

S beam
at 70 lb per foot

2 channels
at 15 lb per foot

29′

38′

27′

Goodheart-Willcox Publisher

38. The following diagram shows the measurements for a frame. Eighty-five frames are to be fabricated. What is the total number of inches of tubing required?

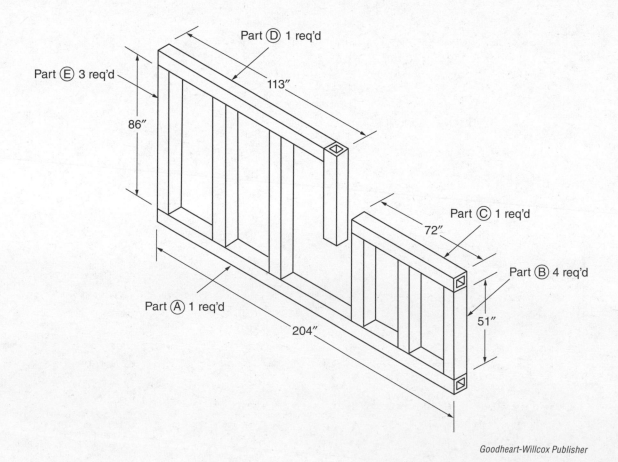

Part Ⓓ 1 req'd

Part Ⓔ 3 req'd

113″

86″

Part Ⓒ 1 req'd

72″

Part Ⓑ 4 req'd

51″

Part Ⓐ 1 req'd

204″

Goodheart-Willcox Publisher

Division of Whole Numbers

Key Terms

dividend   divisor   remainder
division   quotient

Introduction

Division is the process of finding how many times one number can be contained in another number. For example, the number 50 contains the number 10 five times. There are a variety of expressions used to describe this relationship.

> There are 5 tens in 50.
>
> Ten goes into 50 five times.
>
> Fifty divided by 10 equals five.
>
> Ten divides into 50 five times.

While there are a number of ways of expressing this in spoken English, there are also a number of ways of symbolizing division. All of the following symbolize the division of 1,188 by 9.

$$9\overline{)1,188}$$

$$1,188 \div 9$$

$$\frac{1,188}{9}$$

$$1,188/9$$

Method Used to Divide Whole Numbers

To begin, you should know the appropriate terminology. These words (like many math terms) may seem awkward at first. You will see them mainly in math books and occasionally in some formula tables. They are needed to help provide a smooth explanation of the division process.

$$\text{Divisor} \longrightarrow 9\overline{)1,188} \quad \begin{array}{l}132 \longleftarrow \textbf{Quotient} \\ \longleftarrow \textbf{Dividend}\end{array}$$

The *dividend* is the number that will be divided into smaller equal units. It is often the largest number in the equation. The *divisor* is the number of times the dividend will be broken into equal pieces. The *quotient* is the number of equally sized units into which the dividend was broken by the divisor.

1. To begin a division operation, compare the first digit of the dividend with the divisor. Cover all of the dividend, except the leftmost digit, with your finger.

$$7\overline{)131{,}138}$$

2. Then, ask yourself, "Will the divisor go into this digit?" If yes, then ask how many times. If the answer is "No," as in this example, uncover the next digit and ask if the divisor will divide into those digits of the dividend and how often. In this example, the divisor 7 will divide into 13 once. Write the 1 above the 3 as shown. It is extremely important to line up the numbers accurately.

$$\begin{array}{r} 1 \\ 7\overline{)131{,}138} \end{array}$$

3. After writing the 1 above the 3, multiply 7×1 and write the answer under the 13.

$$\begin{array}{r} 1 \\ 7\overline{)131{,}138} \\ 7 \end{array}$$

4. Next, subtract 7 from 13.

$$\begin{array}{r} 1 \\ 7\overline{)131{,}138} \\ \underline{-7} \\ 6 \end{array}$$

5. Borrow, or bring down, the next digit in the dividend. Draw an arrow from that digit in the dividend pointing to the space to the right of the remaining number. Write that digit beside the remaining number. The 6 becomes a 61. This arrow from the dividend to the remaining number shows from where each digit is borrowed in the dividend. These arrows are useful in long division and keep columns neatly aligned.

$$\begin{array}{r} 1 \\ 7\overline{)131{,}138} \\ \underline{-7}\!\!\downarrow \\ 61 \end{array}$$

6. Again, compare the divisor and leftmost dividend digits. "Does the divisor go into 61?" Yes, 7 goes into 61 eight times. Write an 8 directly above the 1 in the dividend. Multiply 7×8 to get 56 and write this product beneath the 61. Perform subtraction.

$$\begin{array}{r} 18 \\ 7\overline{)131{,}138} \\ \underline{-7} \\ 61 \\ \underline{-56} \\ 5 \end{array}$$

7. Continue repeating this cycle with each of the remaining digits.

$$
\begin{array}{r}
18{,}734 \\
7\overline{)131{,}138} \\
\underline{-7} \\
61 \\
\underline{-56} \\
51 \\
\underline{-49} \\
23 \\
\underline{-21} \\
28 \\
\underline{-28} \\
0
\end{array}
$$

Division Equation Variations

The previous example had a dividend and divisor that divided easily. Not all problems proceed so smoothly. There are two variations you will encounter. First, sometimes the divisor cannot divide into the number that has been brought down.

$$
\begin{array}{r}
1 \\
7\overline{)728} \\
\underline{-7} \\
02
\end{array}
$$

In this example, the number 2 has been brought down, but the divisor does not go into 2. In this case, place a zero above the 2 in the dividend and bring down the next number.

$$
\begin{array}{r}
10 \\
7\overline{)728} \\
\underline{-7} \\
028
\end{array}
$$

Forgetting to place this zero in the quotient is an error that a beginner can easily make. Now, proceed with the problem.

$$
\begin{array}{r}
104 \\
7\overline{)728} \\
\underline{-7} \\
028 \\
\underline{-28} \\
0
\end{array}
$$

The second division equation variation is when the divisor does not divide evenly into the dividend. In these cases, you will have a *remainder*. This is a number remaining after final subtraction of divisor and dividend digit products.

$$
\begin{array}{r}
259 \text{ r } 1 \\
5\overline{)1{,}296} \\
\underline{-10} \\
29 \\
\underline{-25} \\
46 \\
\underline{-45} \\
1
\end{array}
$$

The remainder here is 1. Include the remainder with the rest of the quotient but separate it with an *r*. Always check your accuracy.

Checking Division by Multiplying

Check your work by multiplying the quotient by the divisor. If there was a remainder, add it to the product, and the answer should be the same as the dividend. The equation below checks the accuracy of the remainder problem example.

$$
\begin{array}{r}
259 \\
\times\,5 \\
\hline
1{,}295 \\
+\,1 \\
\hline
1{,}296
\end{array}
$$

Do Not Crowd

As you can see, division problems tend to use a lot of space. Always give yourself plenty of room. Do not try to squeeze the figures into a small space. Trying to save space will only increase your chances of making an error.

Denominate Numbers

As in multiplication equations, denominate numbers in division equations do not need to have the same unit. However, the divisor's unit is one part of the combined unit of the dividend. In a division equation of denominate numbers, the divisor's unit of measurement can cancel out the part of the dividend's unit of measurement that they share in common. In this way, compound units consisting of two different units can be separated into their distinct parts.

$$
\begin{array}{r}
25' \\
5\ \text{lb}\ \overline{)\,125\ \text{ft-lb}} \\
-10 \\
\hline
25 \\
-25 \\
\hline
0
\end{array}
$$

In dividing 125 ft-lb by 5 lb, treat the numbers like a normal division equation. However, the units are handled differently. The pound (lb) unit cancels out pound (lb) in the foot-pound (ft-lb) unit, resulting in feet (ft).

Squared units can be viewed as a type of compound unit. In calculating a division equation, a squared unit can be divided into a single unit.

$$
\begin{array}{r}
6'' \\
12\ \text{in}\ \overline{)\,72\ \text{in}^2} \\
-72 \\
\hline
0
\end{array}
$$

In dividing 72 in^2 by 12", treat the numbers like a normal division equation. The inch unit in the divisor cancels one of the inch units in the dividend (in^2 is the same as in-in), resulting in an inch (in) unit for the quotient. These concepts will be reinforced in later units on measurement.

Name _____ **Date** _____ **Class** _____

Divide the following number groupings. Show all your work. Be certain the columns line up. Box your answers.

1. $8\overline{)216}$

2. $29\overline{)348}$

3. $6\overline{)618}$

4. $7\overline{)8,050}$

5. $9\overline{)81,064}$

6. $35\overline{)2,410}$

7. $143\overline{)1,001}$

8. $365\overline{)44,895}$

9. $400\overline{)160,000}$

10. $4,826 \div 3,519 =$

11. 1,840 divided by 72 =

12. $1,066\overline{)981,734}$

13. 18,604/290 =

14. $1,776\overline{)3,528,912}$

15. 245 divided into 6,195 =

16. $8,521\overline{)340,915,264}$

17. How many 335s are there in 6,369?

18. Divide 648 into 8,759.

19. A company that manufactures electric light fixtures used 16,653 lb of solder in one year. If there were 273 working days in the year, how many pounds of solder would be consumed on an average day?

20. Welding a heavy steel plate with a V-groove joint required 1,248 ounces of filler. The plate was 3″ thick and 208″ long. How many ounces of filler were used for each inch of weld?

21. A piece of angle iron 120″ long is to be cut into 19″ pieces. Assuming there is no loss of material due to cutting, how many 19″ long pieces would be obtained?

Use this information to answer the questions that follow. A fabricating shop was successful in its bid for the contract to supply railings for a bridge across the Detroit River. The bridge was to be 4,784' long and railings were to be provided for both sides of the bridge. The railings were welded in 16' sections and delivered to the job site.

22. How many sections of railing were built?

23. If each section weighed 642 lb, what was the total weight of the railings?

24. A supply house ordered 14 cartons of aluminum flux in cans. Each fully packed carton weighed 216 lb (excluding the weight of the carton). If each can weighed 3 lb, how many cans of flux are in the order?

25. During a recent repair of the Golden Gate Bridge, 30,960' of on-site welding was laid. The job was completed in 169 days. What was the average number of feet of weld laid per day? Express your answer to the nearest foot.

26. A job calls for clips to be welded at each end of a 1,863' seawall. Clips are also to be welded 18" apart along the length of the wall. How many clips will be used on this job?

Name _____ **Date** _____ **Class** _____

27. A 96″ piece of flat bar is to have 11 center lines scribed at equal distances apart. How far apart are the center lines? Hint: Draw a sketch of the bar and the center lines.

28. What is the distance between the centers of the evenly spaced holes shown in the following illustration?

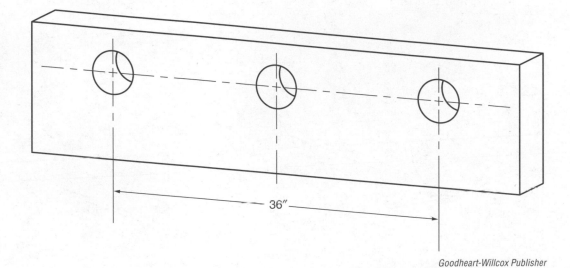

36″

Goodheart-Willcox Publisher

29. What is the distance between the centers of the evenly spaced holes shown in the following illustration?

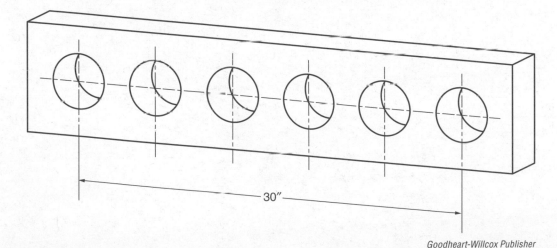

30″

Goodheart-Willcox Publisher

30. The job order for the weldment shown in the following illustration calls for it to have 37 holes drilled evenly spaced apart. Calculate the distance between the holes.

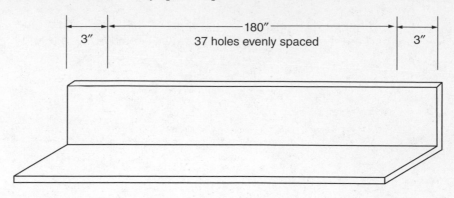

Goodheart-Willcox Publisher

31. How many 17″ pieces can be cut from the 16,761″ roll of steel shown in the following illustration?

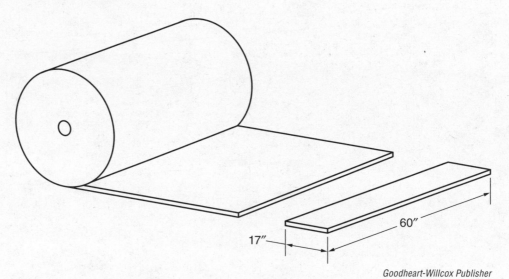

Goodheart-Willcox Publisher

32. Illustrated is one of eight bundles of round bar stock that make up a delivery weighing 14,208 lb. What is the weight of each bar?

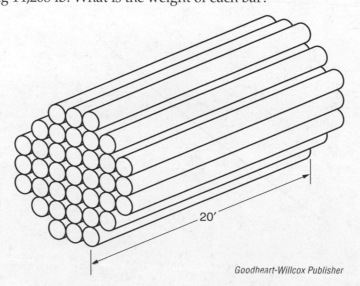

Goodheart-Willcox Publisher

Name _____   **Date** _____   **Class** _____

33. Studs are to be welded to this plate in the following manner. Studs along the length are to be spaced 9″ apart and studs along the width are to be spaced 8″ apart. How many studs are required?

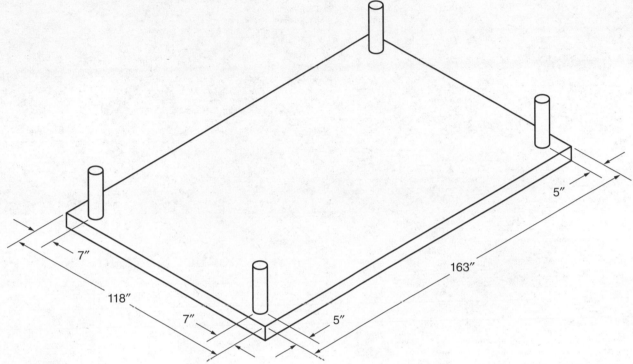

Goodheart-Willcox Publisher

SECTION 2
Common Fractions

Section Objectives

After studying this section, you will be able to:

- Explain the parts of a fraction.
- Provide examples of proper fractions, improper fractions, and mixed numbers.
- Demonstrate how to reduce equivalent fractions.
- Demonstrate how to convert between mixed numbers and improper fractions.
- Perform addition of fractions.
- Perform subtraction of fractions.
- Perform multiplication of fractions.
- Perform division of fractions.

Introduction to Common Fractions

tahkani/Shutterstock.com

Key Terms

common fraction

denominator

equivalent fraction

higher terms

improper fraction

lower terms

lowest terms

mixed number

numerator

proper fraction

reducing

terms

Introduction

When you divide something, such as a steel rod, into a number of equal pieces, each of those pieces can be called a fraction of the whole rod. If the rod is cut into ten equal parts, each part is one-tenth of the whole. This can be expressed in numbers by the fraction 1/10. A fraction such as this is properly called a ***common fraction***. In everyday usage, it is often referred to as a fraction. Three of the parts make up 3/10 of the whole, nine parts would be 9/10, and finally, all ten parts could be written as 10/10. Since all ten parts make up the whole, you can see that 10/10 = 1. Any fraction containing all the parts is equal to 1, such as 64/64, 17/17, 256/256, 1/1, 12/12.

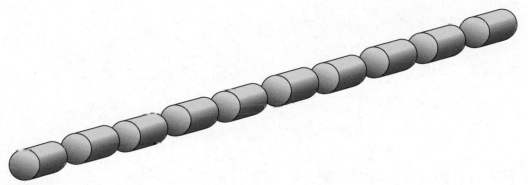

Goodheart-Willcox Publisher

The following three illustrations show fractional parts of an inch.

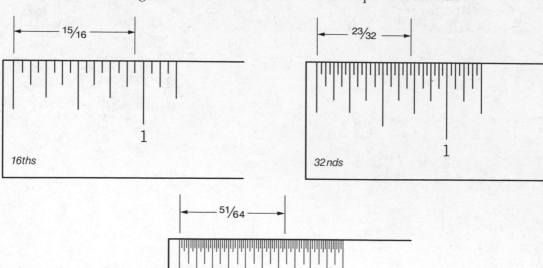

Goodheart-Willcox Publisher

Parts of a Fraction

The parts of a fraction are as follows.

$$\frac{3}{4} \quad \begin{array}{l} \leftarrow \textbf{numerator} \\ \leftarrow \textbf{denominator} \end{array}$$

An easy way to remember these terms and to keep them separate is to think that the *d*enominator is *d*own on the bottom. Associate the *d* in <u>d</u>enominator with the *d* in <u>d</u>own. The numerator and denominator together are referred to as the ***terms*** of the fraction. The ***denominator*** refers to how many equal parts the item has been divided into. The ***numerator*** refers to how many of those parts you are using.

Proper Fractions

A ***proper fraction*** has a numerator smaller than the denominator. Examples are 1/4, 18/57, and 9/32.

Improper Fractions

An ***improper fraction*** has a numerator larger than the denominator. These fractions are "top heavy." Examples are 15/7, 29/17, and 9/8.

Mixed Numbers

A ***mixed number*** consists of a whole number and a fraction. Examples are 1 1/2, 37 11/16, and 108 5/13.

Equivalent Fractions

Equivalent fractions are fractions that are equal in value to each other. Any fraction can be expressed as an equivalent fraction in higher terms or lower terms.

Equivalent Fractions in Higher Terms

To express a fraction in **higher terms**, multiply the numerator and denominator of a fraction by the same number. Since both terms were raised by the same number, the value of the two fractions will be the same, as illustrated:

$$\frac{3 \text{ multiplied by } 7 =}{4 \text{ multiplied by } 7 =} \frac{21}{28}$$

The fractions 3/4 and 21/28 are equivalent fractions.

Equivalent Fractions in Lower Terms

Expressing a fraction in **lower terms** is called **reducing**. If you divide both the numerator and denominator of a fraction by the same number, the fraction will be reduced. The value of the fraction will not be changed. The numbers must divide evenly (that is, without a remainder) as illustrated:

$$\frac{24 \text{ divided by } 6 =}{30 \text{ divided by } 6 =} \frac{4}{5}$$

The fractions 24/30 and 4/5 are equivalent fractions.
Here is another example:

$$\frac{\cancel{\cancel{42}}^{\;6}}{\cancel{\cancel{56}}_{\;8}} = \frac{\cancel{6}^{\,3}}{\cancel{8}_{\,4}} = \frac{3}{4}$$

When a fraction cannot be reduced further, it is considered to be in its **lowest terms**. When you arrive at the final answer, you are expected to reduce the answer to its lowest terms.

Reducing Improper Fractions

Improper fractions are reduced by dividing the numerator by the denominator. If there is a remainder, it becomes the new numerator of the fraction. See the following examples.

$$\frac{19}{4} \qquad\qquad \frac{22}{10} \qquad\qquad \frac{35}{5}$$

$$4\overline{)19} \qquad\qquad 10\overline{)22} \qquad\qquad 5\overline{)35}$$

$$\begin{array}{c} 4 \text{ r } 3 \\ 4\overline{)19} \\ -16 \\ \hline 3 \end{array} \qquad \begin{array}{c} 2 \text{ r } 2 \\ 10\overline{)22} \\ -20 \\ \hline 2 \end{array} \qquad \begin{array}{c} 7 \\ 5\overline{)35} \\ -35 \\ \hline 0 \end{array}$$

$$4\frac{3}{4} \qquad\qquad 2\frac{1}{5} \qquad\qquad 7$$

Changing Mixed Numbers to Improper Fractions

This is an operation useful in solving certain math problems you will encounter. For this example, we will use the mixed number 7 2/3.

1. Multiply the whole number by the denominator.

$$7 \times 3 = 21$$

2. Add the numerator to the value of the whole number and denominator product.

$$21 + 2 = 23$$

3. Place this sum over the denominator of the original mixed number.

$$7\,\frac{2}{3} = \frac{23}{3}$$

Unit 6 Practice

Name _____ Date _____ Class _____

Express the following as equivalent fractions. Show all your work.
Box your answers.

1. $\dfrac{5}{8} = \dfrac{?}{16}$

2. $\dfrac{1}{2} = \dfrac{?}{200}$

3. $\dfrac{1}{1} = \dfrac{?}{98}$

4. $\dfrac{20}{64} = \dfrac{?}{16}$

5. $\dfrac{28}{32} = \dfrac{?}{8}$

6. $\dfrac{128}{256} = \dfrac{?}{2}$

7. $\dfrac{11}{13} = \dfrac{?}{65}$

8. $\dfrac{3}{19} = \dfrac{?}{133}$

9. $\dfrac{4}{5} = \dfrac{?}{100}$

Determine whether the following pairs of fractions are equivalent fractions. Answer each group with *yes* or *no*.

10. $\dfrac{2}{3}$ and $\dfrac{65}{99}$

11. $\dfrac{3}{4}$ and $\dfrac{21}{28}$

12. $\dfrac{17}{64}$ and $\dfrac{17}{128}$

13. $\dfrac{1}{1}$ and $\dfrac{2}{1}$

14. $\dfrac{10}{11}$ and $\dfrac{110}{121}$

15. $\dfrac{19}{33}$ and $\dfrac{76}{132}$

16. $\dfrac{11}{111}$ and $\dfrac{99}{999}$

17. $\dfrac{33}{19}$ and $\dfrac{132}{76}$

18. $\dfrac{100}{10}$ and $\dfrac{10}{100}$

Reduce the following improper fractions. Show all your work. Box your answers.

19. $\dfrac{18}{4}$

20. $\dfrac{111}{11}$

21. $\dfrac{141}{8}$

22. $\dfrac{1968}{16}$

23. $\dfrac{7}{2}$

24. $\dfrac{75}{25}$

25. $\dfrac{1505}{150}$

26. $\dfrac{10830}{833}$

27. $\dfrac{2}{1}$

Change the following mixed numbers to improper fractions. Show all your work.
Box your answers.

28. $1 \dfrac{1}{2}$

29. $12 \dfrac{3}{4}$

30. $17 \dfrac{11}{64}$

31. $29 \dfrac{1}{8}$

32. $37 \dfrac{3}{7}$

33. $185 \dfrac{5}{9}$

34. $13 \dfrac{11}{13}$

35. $10 \dfrac{10}{10}$

36. $7 \dfrac{7}{9}$

Express the following fractions in the lowest terms. Show all your work.
Box your answers.

37. $\dfrac{2}{4}$

38. $\dfrac{30}{36}$

39. $\dfrac{98}{112}$

40. $\dfrac{12}{108}$

41. $\dfrac{64}{656}$

42. $\dfrac{16}{2048}$

43. $\dfrac{39}{65}$

44. $\dfrac{119}{1700}$

45. $\dfrac{505}{1000}$

Addition of Fractions

Key Terms

common denominator
lowest common denominator (LCD)

Introduction

The basic operations of addition, subtraction, multiplication, and division are performed with fractions and combinations of fractions and whole numbers. Performing these math calculations is a fundamental task of a welder.

Methods Used to Add Fractions

Three main situations are encountered when adding fractions. Fractions with common denominators, fractions without common denominators, and mixed numbers without common denominators can be added.

Adding Fractions with Common Denominators

Fractions can be added only if they have the *same* denominators. When two fractions have the same denominator, they share a **common denominator**. For example, 3/11 and 4/11 have a common denominator. To add these fractions, just add the numerators (3 + 4 = 7) and place this sum over the common denominator (7/11).

Adding Fractions without Common Denominators

Fractions without common denominators can be added by first changing the denominators so they are the same. The whole operation is easier if the denominator selected is the lowest one possible. Your first objective, then, is to find the **lowest common denominator (LCD)**. An efficient way of doing this is as follows:

1. Review the fractions that are to be added.

$$\frac{5}{6} + \frac{3}{8} + \frac{2}{3}$$

2. In a blank space, write all the denominators in a row. It is a good idea to separate them by commas so they do not blend together.

$$6, 8, 3$$

3. Find a number that divides evenly into at least two of the numbers listed. In this example, 2 will work. Divide using the format illustrated. Record the results of the division.

$$2\,\underline{|6, 8, 3}$$
$$3, 4, 3$$

Two divides evenly into 6 three times, so 3 is written below the 6. Two divides evenly into 8 four times, so 4 is written below the 8. Since 2 does not divide evenly into 3, the 3 is brought down.

4. This step follows the pattern of the previous steps. This time, inspect the new line of 3, 4, 3 to find a number that divides evenly into at least two of the numbers. In this case, the number is 3.

$$2\,\underline{|6, 8, 3}$$
$$3\,\underline{|3, 4, 3}$$
$$1, 4, 1$$

Since 3 does not divide evenly into 4, the 4 is brought down. At this point, the line of denominator digits cannot be reduced further.

5. Multiply the line of reduced denominators (1, 4, 1) and the numbers by which they were divided (2, 3):

$$1 \times 4 \times 1 \times 2 \times 3 = 24$$

The LCD is 24.

6. Express each fraction from the original equation as an equivalent fraction with a denominator of 24. To review expressing equivalent fractions in higher terms, see Unit 6.

$$\frac{5}{6} + \frac{3}{8} + \frac{2}{3} = \frac{20}{24} + \frac{9}{24} + \frac{16}{24}$$

7. Add the numerators of the fractions.

$$\frac{20}{24} + \frac{9}{24} + \frac{16}{24} = \frac{45}{24}$$

8. If necessary, change to a mixed number and reduce the fraction to its lowest terms.

$$\frac{45}{24} = 1\frac{21}{24} = 1\frac{7}{8}$$

Adding Mixed Numbers without Common Denominators

There are several ways to add mixed numbers without common denominators. The most common method is explained here.

1. Add $2\frac{3}{4} + 5\frac{1}{7}$

2. Rewrite the question in this manner.

$$2 \, \frac{3}{4}$$

$$+ \; 5 \, \frac{1}{7}$$

3. Find the LCD and write as equivalent fractions.

$$2 \, \frac{21}{28}$$

$$+ \; 5 \, \frac{4}{28}$$

4. Add the whole numbers together and then add the fractional parts together.

$$2 \, \frac{21}{28}$$

$$+ \; 5 \, \frac{4}{28}$$

$$7 \, \frac{25}{28}$$

5. Reduce the answer to its lowest terms, if possible.

Since 7 25/28 is already the lowest term in this case, no further operations are necessary.

Work Space/Notes

Name _____ **Date** _____ **Class** _____

Solve the following equations. Show all your work. Box your answers.

1. $\dfrac{3}{18} + \dfrac{7}{18} + \dfrac{1}{18} =$

2. $\dfrac{21}{96} + \dfrac{13}{96} + \dfrac{37}{96} =$

3. $\dfrac{109}{219} + \dfrac{25}{219} + \dfrac{34}{219} =$

4. $\dfrac{3}{5} + \dfrac{4}{15} =$

5. $\dfrac{17}{25} + \dfrac{4}{5} + \dfrac{49}{50} =$

6. $\dfrac{15}{28} + \dfrac{3}{4} + \dfrac{6}{7} =$

7. $\dfrac{3}{4} + \dfrac{1}{3} + \dfrac{4}{5} + \dfrac{1}{2} =$

8. $\dfrac{3}{8} + \dfrac{3}{7} + \dfrac{7}{12} =$

9. $\dfrac{107}{120} + \dfrac{85}{150} =$

10. $\dfrac{7}{27} + \dfrac{31}{36} + \dfrac{3}{23} =$

11. $\dfrac{7}{64} + \dfrac{41}{88} + \dfrac{53}{92} =$

12. $\dfrac{117}{365} + \dfrac{99}{200} + \dfrac{87}{260} =$

13. $13\dfrac{1}{4} + 5\dfrac{5}{8} + 4\dfrac{9}{16} =$

14. $\dfrac{3}{56} + \dfrac{3}{16} + \dfrac{7}{28} + \dfrac{2}{7} =$

15. $38\dfrac{19}{34} + 119\dfrac{11}{12} =$

16. $11\dfrac{1}{4} + 27\dfrac{3}{16} + \dfrac{7}{8} + 10\dfrac{1}{2} =$

17. $\dfrac{25}{140} + \dfrac{9}{204} =$

18. $1\dfrac{109}{187} + \dfrac{75}{121} =$

19. According to their time cards, the following people spent the indicated amount of time on Job #8815. What was the total amount of work time spent on the job?

Job #8815	
Name	**Hours**
Lee Jones	44 1/4
Jamil El Helou	185 3/4
Filomena Ferrari	121 1/2
Pat Anderson	15
Luis Martin	29 1/4
Guy Laframbois	6 1/2

Goodheart-Willcox Publisher

20. A stainless steel rod, 1/2" in diameter, has been cut into five pieces. According to the lengths of each of the five pieces listed below, what was the original length of the rod? Ignore losses due to cutting.

    Twenty-one and three-quarter inches
    Seven inches
    Thirty-two and seven-thirty-seconds inches
    Eleven and one-half inches
    Sixteen and fifteen-sixteenth inches

21. At Camelot Metal Products, a welder has to cut a 1" square rod into three pieces of the following lengths—21 3/8", 26 1/4", and 17 5/16". Each cut will waste 1/4" of material. The pieces are cut from a 96" rod. What length of the rod is used to make the three pieces?

22. What is the overall length of a machine part consisting of five pieces measuring 7/8", 41 9/16", 12 27/32", 13/64", and 1 1/2"?

23. Leamington Brass produced a machined brass tube with an inside diameter of 21 7/8" and a thickness of 11/32". What is the outside diameter?

Name _____   **Date** _____   **Class** _____

24. The individual parts of a weldment had the following weights—135 1/2 lb, 19 3/16 lb, 71 1/4 lb, 6 3/8 lb, and 1 1/4 lb. The filler material used in welding added an additional 2 3/32 lb. What is the final weight of the completed weldment?

25. Find the distance from stud Ⓐ to stud Ⓑ.

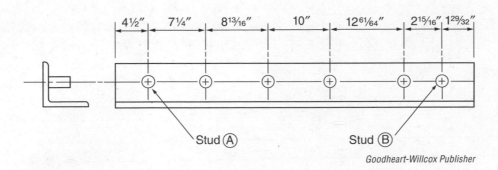

Goodheart-Willcox Publisher

26. Calculate the overall length of the assembly in the following diagram.

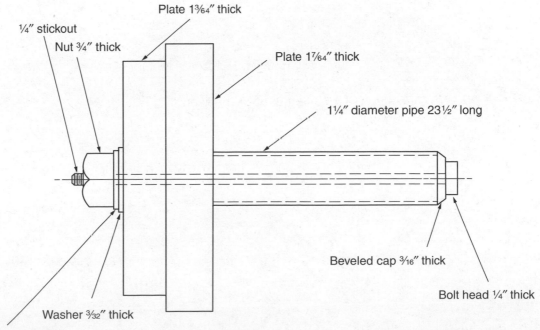

Review the measurements of the three objects in the following diagram to answer the following three questions.

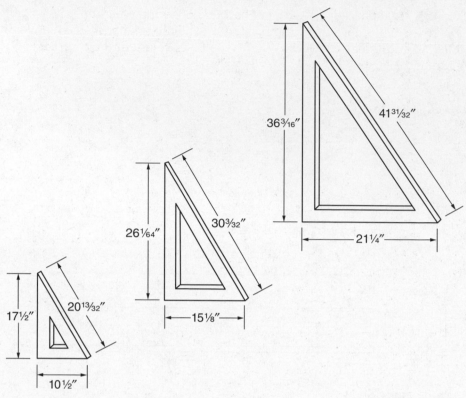

27. Calculate the total length of the vertical sides.

28. Calculate the total length of the horizontal sides.

29. Calculate the total length of the angular sides.

Name _____   **Date** _____   **Class** _____

30. Calculate the overall length of this test bar after welding.

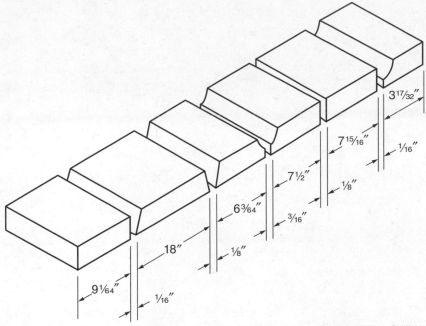

Goodheart-Willcox Publisher

Review the following diagram to answer the following two questions.

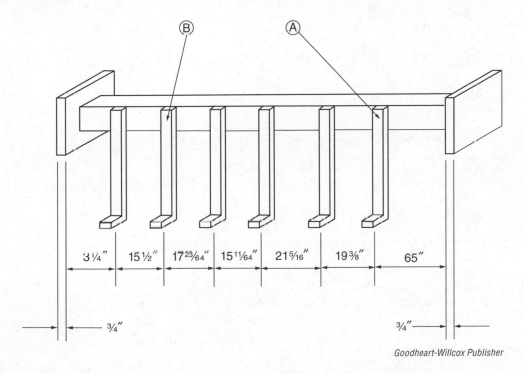

Goodheart-Willcox Publisher

31. What is the distance from hanger Ⓐ to hanger Ⓑ?

32. What is the total length of the weldment in the previous illustration?

33. What is the total length of angle iron used in welding the frame shown in the following illustration?

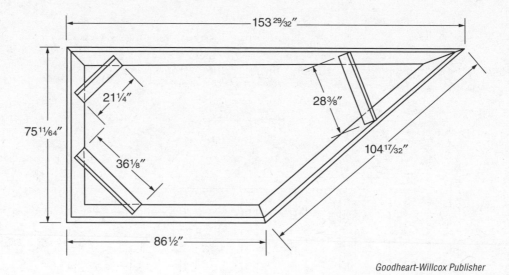

Goodheart-Willcox Publisher

UNIT 8

Subtraction of Fractions

Introduction

Subtracting fractions is similar to adding fractions. The method is generally the same. You will work to find fractions with common denominators before subtracting the numerators.

Methods Used to Subtract Fractions

Three main situations are encountered when subtracting fractions. Fractions with common denominators, fractions without common denominators, and mixed numbers without common denominators can be subtracted.

Subtracting Fractions with Common Denominators

Fractions with common denominators are easily subtracted. Simply subtract one numerator from the other numerator and place the result over the common denominator.

$$\frac{9}{17} - \frac{3}{17} = \frac{6}{17}$$

Subtracting Fractions without Common Denominators

To subtract fractions without common denominators, first change the denominators so they are the same. As explained in Unit 7, finding the LCD is usually the easiest way to do this. Review that explanation, if necessary. Then study the following example:

$$\frac{13}{14} - \frac{6}{63} -$$

$$7\overline{)14, 63}$$
$$\quad 2, \quad 9$$

$7 \times 2 \times 9 = 126$

LCD = 126

Therefore:

$$\frac{13}{14} - \frac{6}{63} = \frac{117}{126} - \frac{12}{126} = \frac{105}{126}$$

Subtracting Mixed Numbers without Common Denominators

To subtract mixed numbers without common denominators, first change the fractional parts to common denominators. Then subtract the fractional parts. Finally, subtract the whole numbers.

$$17 \frac{3}{4} = 17 \frac{21}{28}$$

$$-4 \frac{1}{7} = -4 \frac{4}{28}$$

$$13 \frac{17}{28}$$

Borrowing a Whole Number for a Fraction

Not all subtraction equations follow the exact steps listed above. A complication occurs when the fractional part to be subtracted (the subtrahend) is larger than the fraction from which it is being subtracted (the minuend).

$$11 \frac{9}{16}$$

$$-5 \frac{12}{16}$$

When this situation occurs, borrow 1 from the whole number and add it to the fraction part of the number.

$$1\overset{0}{\cancel{1}} \frac{9}{16} \left(+ \frac{16}{16} \right)$$

$$-5 \frac{12}{16}$$

When a fraction borrows from a whole number, the whole number must be changed into a fraction before it can be added to the fraction. When a 1 is changed to a fraction, it has the same number in the numerator as it has in the denominator. Examples of how the number 1 appears as a fraction: 3/3, 8/8, 16/16. For review, see Unit 6.

$$10 \frac{25}{16}$$

$$-5 \frac{12}{16}$$

$$5 \frac{13}{16}$$

After reducing the whole number from the borrowing and calculating the new numerator, the problem can be solved by subtracting each column right to left.

Name _____ **Date** _____ **Class** _____

Solve the following equations. Reduce to lowest terms. Show all your work. Box your answers.

1. $\dfrac{5}{9} - \dfrac{2}{9} =$

2. $\dfrac{13}{16} - \dfrac{5}{16} =$

3. $\dfrac{62}{101} - \dfrac{31}{101} =$

4. $\dfrac{7}{8} - \dfrac{1}{4} =$

5. $\dfrac{15}{16} - \dfrac{7}{8} =$

6. $35\dfrac{17}{20} - 25\dfrac{11}{17} =$

7. Reduce 738 7/32 by 48 9/32.

8. Find the difference between 6,499 3/4 and 6,460 1/2.

9. 111/119 less 111/238.

10. Subtract 19/64 from 5/8.

11. Take 49 2/3 away from 65.

12. Calculate the difference between 2,020 1/2 and 1,100 1/2.

13. $7\dfrac{1}{3} - 6\dfrac{2}{3} =$

14. $11\dfrac{9}{16} - \dfrac{7}{8} =$

15. $1\dfrac{3}{10} - \dfrac{5}{6} =$

16. $19\dfrac{2}{5} - 18\dfrac{9}{10} =$

17. $45\dfrac{1}{3} - 32\dfrac{2}{3} =$

18. $24\dfrac{1}{4} - 15\dfrac{2}{3} =$

19. A piece of steel (US Standard gage #28) 55 5/8″ long was sheared from a sheet measuring 118 7/16″ long. What length of the original sheet remains?

20. The sides of a square piece of steel measure 18 3/8″. Using two cuts, the piece is sheared to 11 7/64″ × 13 3/16″. What is the width of each removed piece?

Use this information to answer questions 21–23. A length of Minnesota pipeline measuring 87 1/2 miles is due for inspection. The contract went to Twin City Consulting, Inc. The company estimated they could inspect 5 1/8 miles of line per week.

21. What length of pipeline would remain to be inspected after one week?

22. What length of pipeline would remain to be inspected after two weeks?

23. What length of pipeline would remain to be inspected after three weeks?

24. In Detroit's Industrial Softball League, halfway through the season the Pipefitters were 17 games behind the leader and ranked in 5th place. The Boilermakers were in 4th place and 13 1/2 games behind the leader. How many games behind the Boilermakers were the Pipefitters?

25. The wall thickness of a piece of round tubing is 3/16″ and the outside diameter is 3 1/4″. What is the inside diameter?

Name _____   **Date** _____   **Class** _____

26. Four pieces of square tubing measuring 22 5/32", 17", 20 1/2", and 21 1/4" are to be cut from a stock piece of 1" × 1" square tubing that is 131 1/4" long. Each saw cut wastes 3/32" of material. What will be the length of the stock piece after cutting all four pieces?

27. If the 1/4" square bar in the following illustration is reduced to 20 1/4", how long would the piece removed be? The saw cut will waste 1/16" of material.

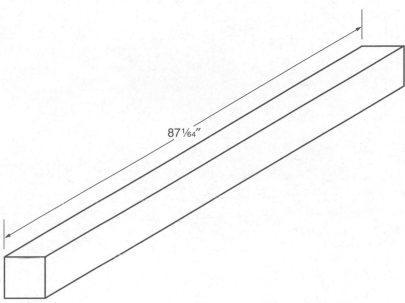

87 1/64"

Goodheart-Willcox Publisher

28. If the 1 1/4" × 1 1/4" square tubing in the following illustration is reduced to 23 5/8", how long would the piece removed be? The saw cut will waste 1/16" of material.

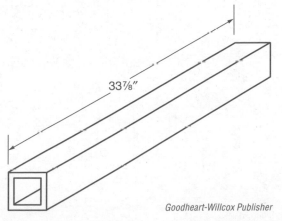

33 7/8"

Goodheart-Willcox Publisher

29. If the 2″ channel in the following illustration is reduced to 40 5/8″, how long would the piece removed be? The saw cut will waste 1/16″ of material.

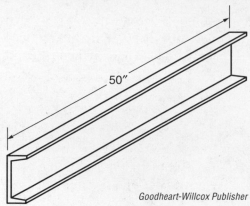

50″

Goodheart-Willcox Publisher

30. If the 3/4″ round bar in the following illustration is reduced to 38 1/4″, how long would the piece removed be? The saw cut will waste 1/16″ of material.

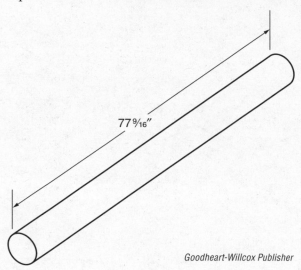

77 9/16″

Goodheart-Willcox Publisher

31. In the following diagram, calculate the length of Ⓐ.

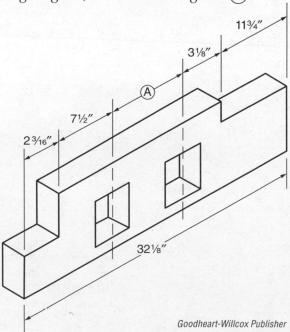

11¾″

3⅛″

Ⓐ

7½″

2 3/16″

32⅛″

Goodheart-Willcox Publisher

Name _____   **Date** _____   **Class** _____

Refer to the following diagram to answer questions 32–34.

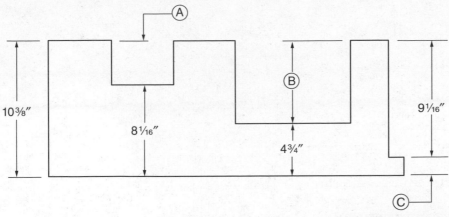

Goodheart-Willcox Publisher

32. Find the length of Ⓐ as shown in the diagram.

33. Find the length of Ⓑ as shown in the diagram.

34. Find the length of Ⓒ as shown in the diagram.

Refer to the following diagram to answer questions 35–36.

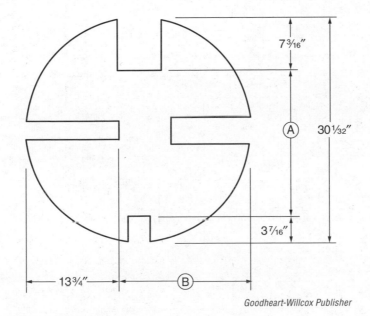

Goodheart-Willcox Publisher

35. Calculate the length of Ⓐ.

36. Calculate the length of Ⓑ.

37. Find the length of (A) in the following diagram.

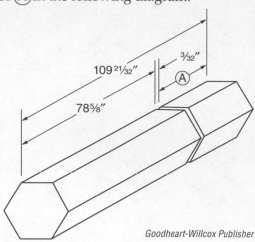

Goodheart-Willcox Publisher

38. Three pieces measuring 18 11/64", 9 3/8", and 82 1/16" will be cut from this half-round stock. Each saw cut wastes 3/32" of material. What will be the length of the remaining piece?

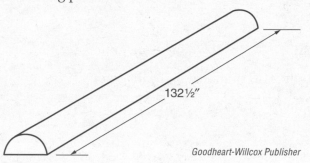

Goodheart-Willcox Publisher

Review the following diagram to answer the following questions.

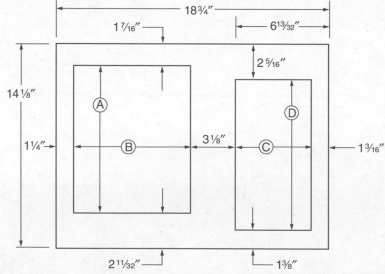

Goodheart-Willcox Publisher

39. Calculate the length of (A).

40. Calculate the length of (B).

41. Calculate the length of (C).

42. Calculate the length of (D).

Multiplication of Fractions

tahkani/Shutterstock.com

Introduction

It may seem surprising, but multiplying fractions is easier than adding or subtracting fractions. This is because the denominators do not have to be the same. You do not have to find a common denominator.

Three expressions are commonly used to indicate multiplication of fractions.

$$\frac{1}{5} \times \frac{3}{4}$$

$$\frac{1}{5} \text{ times } \frac{3}{4}$$

$$\frac{1}{5} \text{ of } \frac{3}{4}$$

The one expression that may seem peculiar is *of*; however, through common usage, this word has come to mean multiplication.

Methods Used to Multiply Fractions

To multiply $3/4 \times 5/7$, multiply the two numerators and then multiply the two denominators.

$$\frac{3}{4} \times \frac{5}{7} = \frac{15}{28}$$

As always, reduce the answer to the lowest terms. Sometimes a question can be reduced to a simpler form before beginning the multiplication. You can do this by finding a number that divides evenly into *any* of the numerators and denominators. To better understand this, you may want to rewrite the equation as a visual cue.

$$\frac{16}{100} \times \frac{25}{37} = \frac{16 \times 25}{100 \times 37}$$

Arranging a fraction multiplication equation as a single fraction with one numerator and one denominator should remind you of the flexibility allowed in reducing the equation. Now, begin reducing the terms.

$$\frac{16 \times \overset{1}{\cancel{25}}}{\underset{4}{\cancel{100}} \times 37} =$$

Always look to reduce to the lowest terms possible. Even after reducing the 25 and 100, this equation can be reduced further.

$$\frac{\overset{4}{\cancel{16}} \times \overset{1}{\cancel{25}}}{\underset{\underset{1}{\cancel{4}}}{\cancel{100}} \times 37} =$$

Continue doing this until the numerators and denominators cannot be reduced any further. Then perform the multiplication.

$$\frac{\overset{4}{\cancel{16}} \times \overset{1}{\cancel{25}}}{\underset{\underset{1}{\cancel{4}}}{\cancel{100}} \times 37} = \frac{4}{37}$$

You will find that multiplying fractions is much easier if you first reduce to the lowest possible terms.

Multiplying Mixed Numbers

To multiply mixed numbers, first change them to improper fractions. Once the multiplication is done, convert the product back into a mixed number, as shown here.

$$3\frac{1}{2} \times 9\frac{2}{5} = \frac{7}{2} \times \frac{47}{5} = \frac{329}{10} = 32\frac{9}{10}$$

If the task is to multiply a mixed number and a whole number, rewrite the whole number as a fraction with a denominator of 1. For example, 42 is rewritten as 42/1. A typical problem would look like this:

$$42 \times 2\frac{4}{7} =$$

$$\frac{42}{1} \times \frac{18}{7} =$$

$$\frac{\overset{6}{\cancel{42}} \times 18}{1 \times \underset{1}{\cancel{7}}} =$$

$$\frac{6 \times 18}{1 \times 1} = \frac{108}{1} = 108$$

Multiplying More Than Two Fractions

Multiplying more than two fractions follows the same routine as multiplying two fractions. First, try to reduce the question to a simpler form and then multiply the numerators and denominators.

$$\frac{1}{2} \times \frac{\overset{1}{\cancel{8}}}{4} \times \frac{5}{\underset{2}{\cancel{6}}}$$

$$\frac{1}{2} \times \frac{1}{4} \times \frac{5}{2} = \frac{5}{16}$$

Multiplying More Than Two Mixed Numbers

When an equation is written to multiply more than two mixed numbers, first convert the mixed numbers into improper fractions.

$$5\frac{1}{2} \times 4\frac{2}{3} \times 9\frac{3}{5} = \frac{11}{2} \times \frac{14}{3} \times \frac{48}{5}$$

Next, compare numerators and denominators. Reduce any fractions to a simpler equivalent form.

$$\frac{11}{\overset{\cancel{2}}{1}} \times \frac{\overset{7}{\cancel{14}}}{\underset{1}{\cancel{3}}} \times \frac{\overset{16}{\cancel{48}}}{5}$$

Now calculate the equation by multiplying the numerators. Then multiply the denominators.

$$\frac{11}{1} \times \frac{7}{1} \times \frac{16}{5} = \frac{1,232}{5}$$

The resulting improper fraction can be turned into a mixed number by dividing the numerator by the denominator. The quotient becomes the whole number. The divisor becomes the denominator. The remainder becomes the numerator.

$$246 \text{ r } 2 = 246\frac{2}{5}$$

$$
\begin{array}{r}
5\,\overline{)1{,}232} \\
-10 \\
\hline
23 \\
-20 \\
\hline
32 \\
-30 \\
\hline
2
\end{array}
$$

Work Space/Notes

Name _____ **Date** _____ **Class** _____

Solve the following equations. Show all your work. Box your answers.

1. $\dfrac{1}{5} \times \dfrac{1}{3} =$

2. $\dfrac{3}{4} \times \dfrac{7}{15} =$

3. $\dfrac{15}{33} \times \dfrac{12}{61} =$

4. $\dfrac{1}{2} \times \dfrac{1}{2} =$

5. $\dfrac{13}{64} \times \dfrac{8}{9} =$

6. $\dfrac{7}{35} \times \dfrac{12}{144} =$

7. $4 \times \dfrac{3}{16} =$

8. $3\dfrac{1}{3} \times 5\dfrac{1}{2} =$

9. $17\dfrac{2}{5} \times 2\dfrac{1}{4} =$

10. $8\dfrac{1}{2} \times \dfrac{15}{45} =$

11. $9\dfrac{3}{8} \times 9\dfrac{3}{8} =$

12. $15 \times \dfrac{15}{16} =$

13. $\dfrac{16}{27} \times \dfrac{3}{4} \times \dfrac{5}{7} =$

14. $3\dfrac{1}{2} \times 4\dfrac{1}{6} \times 1\dfrac{3}{10} =$

15. $11\dfrac{3}{16} \times 1\dfrac{1}{8} \times 2 =$

16. $\dfrac{23}{140} \times \dfrac{70}{92} \times \dfrac{17}{33} =$

17. $7 \times \dfrac{1}{2} \times \dfrac{3}{4} =$

18. $9 \times \dfrac{4}{15} \times 9\dfrac{4}{15} =$

19. A pipeline is laid at the rate of 1/8 mile per day. How many miles of line would be completed in 29 1/2 days?

20. One cubic foot of water weighs 62 1/2 lb. Find the weight of the contents of a welded steel tank containing 28 cubic feet of water.

21. Last year, Solar Supplies lost 133 1/2 man hours of labor due to accidents. This year, they lost 1 1/2 times that amount. How many hours were lost this year?

22. Kilowatts are changed to horsepower by multiplying the number of kilowatts by 1 1/3. Change 24 1/2 kilowatts to horsepower.

23. Average truck drivers can move their foot from the accelerator to the brake in 13/16 of a second. At 60 miles per hour a truck is traveling 88 feet per second. How far will the truck travel before the driver's foot is moved to the brake?

Name _____   **Date** _____   **Class** _____

Use this information to answer the two questions that follow. Three years ago, YKY Fasteners International produced 5,960,000 bolts of various sizes. The production two years ago was 1 3/5 times the total from three years ago. Last year's production was 1 13/16 times the total from three years ago.

24. How many bolts were produced two years ago?

25. How many bolts were produced last year?

26. Two grain hoppers were built by a crew consisting of three welders who worked on the job 4 1/4 hours a day for 17 days. How many hours of work were spent on the hoppers?

27. A GMAW welder running at a speed of 16 3/8" per minute requires 14 1/2 minutes to complete a V-groove weld on an aluminum sill. What is the length of the sill?

Use this information to answer the questions 28 through 30. During each hour, a shop uses the following quantities of weld material:

132 3/8 lb of 3/16″ wire

81 5/16 lb of 5/64″ wire

25 3/4 lb of 1/4″ wire

28. How many pounds of 1/4″ wire are used in 5 1/6 hours?

29. How many pounds of 3/16″ wire are used in 8 1/2 hours?

30. How many pounds of 5/64″ wire are used in 3 7/60 hours?

31. The following weldment requires 6 2/3 rods. If 40 1/2 weldments are completed in 2 1/4 days, how many rods will be required?

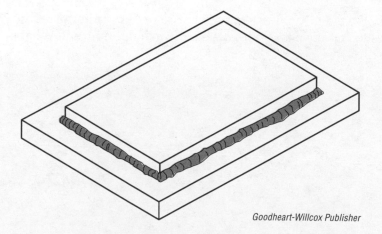

Goodheart-Willcox Publisher

Name _____   **Date** _____   **Class** _____

32. Review the following diagram. All notches are 2 11/16″ wide. All contact surfaces are 3 3/8″ wide. What is the overall length of the rack?

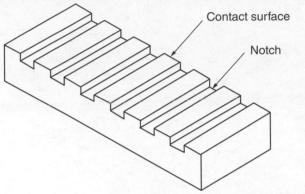

Contact surface

Notch

Goodheart-Willcox Publisher

33. Review the following diagram. Nine of these pillars are to be fabricated. For the purpose of weight reduction, each pillar is to have 13 holes flame cut in the center plate. Seven large holes are to be cut, reducing the weight by 32 19/32 lb per hole. Six smaller holes are to be cut, reducing the weight by 23 7/16 lb per hole. What is the total weight reduction for the entire project?

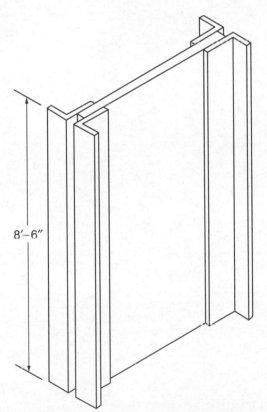

8′–6″

Goodheart-Willcox Publisher

Use this information and the following diagram to answer questions 34 and 35. A flame cutting job using Pattern #1 resulted in 3/16 lb of scrap per part. The scrap rate for Pattern #2 was 2/5 of the rate for Pattern #1.

34. What weight of scrap would result if Pattern #1 is used to produce 5,040 parts?

35. What weight of scrap would result if Pattern #2 is used to produce the same number of parts?

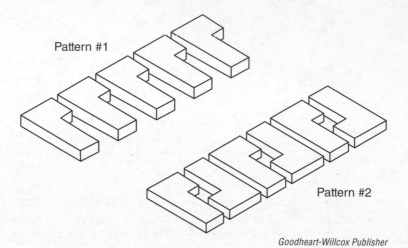

Pattern #1

Pattern #2

Goodheart-Willcox Publisher

Use this information and the following diagram to answer questions 36 and 37. One part, as illustrated, is produced every 47/60 of a minute. One weld bead is used to join each vertical length of the top piece to the lower piece.

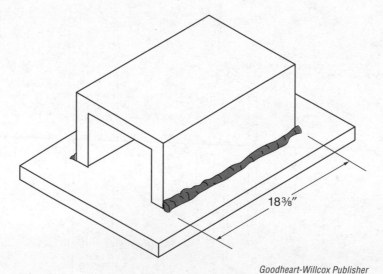

18⅜″

Goodheart-Willcox Publisher

36. How many minutes are required to produce 2,864 parts?

37. What is the total length of weld deposited for 2,864 parts?

Name _____   **Date** _____   **Class** _____

Use this information and the following diagram to answer questions 38 and 39.
The structural work for a new shopping mall includes 1,047 hangers.

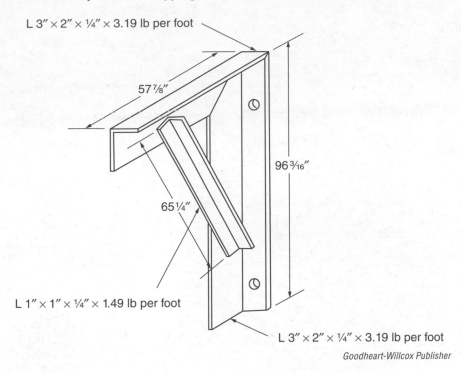

L 3″ × 2″ × ¼″ × 3.19 lb per foot

57⅞″

96³⁄₁₆″

65¼″

L 1″ × 1″ × ¼″ × 1.49 lb per foot

L 3″ × 2″ × ¼″ × 3.19 lb per foot

Goodheart-Willcox Publisher

38. What is the total length of 3″ × 2″ angle used?

39. What is the total length of 1″ × 1″ angle used?

40. Refer to the following diagram. Calculate the weight of the weldment.

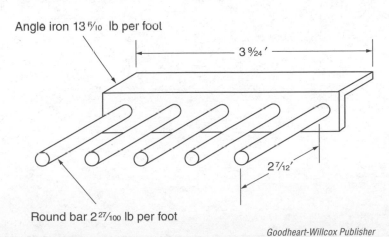

Angle iron 13 ⁶⁄₁₀ lb per foot

3 ⁹⁄₂₄′

2⁷⁄₁₂′

Round bar 2²⁷⁄₁₀₀ lb per foot

Goodheart-Willcox Publisher

Work Space/Notes

UNIT 10

Division of Fractions

Key Terms

complex fraction
invert the divisor

Introduction

Learning to divide fractions is quite easy once you have learned to multiply fractions. You will see why as you read this unit.

Methods Used to Divide Fractions

Fractional division equations closely resemble fractional multiplication equations. In both cases, the fractions are written vertically and separated by the operational math symbol. To divide fractions, *invert the divisor* (that is, turn the divisor upside down by switching the numerator and denominator) and change the operation symbol from division to multiplication.

$$\frac{3}{5} \div \frac{5}{8} = \frac{3}{5} \times \frac{8}{5} = \frac{24}{25}$$

Remember, the divisor is the number doing the division. In this example, 3/5 (the dividend) is being divided by 5/8, the divisor. When dividing fractions, be certain to identify the divisor. If the wrong fraction is inverted, your answer will be incorrect.

It may seem odd that a division problem can suddenly be switched to a multiplication problem. Although they are opposite operations, multiplication and division are closely related. Because of this close relationship, it is possible to convert fractional division problems to multiplication problems as shown above.

Dividing Mixed Numbers

As in multiplying fractions, first change the mixed numbers to improper fractions. Then, invert the divisor and multiply. As usual, try to reduce the problem to a simpler form before proceeding with the multiplication part of the problem. Here is an example:

$$7\frac{3}{4} \div 7\frac{1}{2} =$$

$$\frac{31}{4} \div \frac{15}{2} =$$

$$\frac{31}{\overset{2}{\cancel{4}}} \times \frac{\overset{1}{\cancel{2}}}{15} = \frac{31}{30} = 1\frac{1}{30}$$

Dividing Complex Fractions

There are several ways to symbolize division of fractions. Here are two different formats you can expect to encounter:

The standard format:

$$\frac{5}{8} \div \frac{2}{3}$$

The complex fraction format:

$$\frac{\dfrac{5}{8}}{\dfrac{2}{3}}$$

A *complex fraction* has a fraction for the numerator, a fraction for the denominator, or fractions for both the numerator and denominator. To easily perform this division, rewrite it in the standard format and proceed in the usual manner:

$$\frac{\dfrac{5}{8}}{\dfrac{2}{3}} = \frac{5}{8} \div \frac{2}{3} = \frac{5}{8} \times \frac{3}{2} = \frac{15}{16}$$

Even though a complex fraction may at first appear difficult to divide, it really is quite simple. The same general procedure is followed when the numerator or denominator or both contain mixed numbers. However, in such cases, the mixed number is converted into an improper fraction.

$$\frac{\dfrac{9}{144}}{6\dfrac{3}{4}} = \frac{9}{144} \div 6\frac{3}{4} = \frac{9}{144} \div \frac{27}{4} = \frac{9}{144} \times \frac{4}{27} = \frac{\overset{1}{\cancel{9}}}{\underset{36}{\cancel{144}}} \times \frac{\overset{1}{\cancel{4}}}{\underset{3}{\cancel{27}}} = \frac{1}{108}$$

Name _____ **Date** _____ **Class** _____

Solve the following equations. Show all your work. Box your answers.

1. $\dfrac{1}{3} \div \dfrac{5}{16} =$

2. $\dfrac{4}{5} \div \dfrac{9}{10} =$

3. $\dfrac{5}{6} \div \dfrac{7}{8} =$

4. $\dfrac{7}{11} \div \dfrac{7}{11} =$

5. $\dfrac{3}{4} \div \dfrac{4}{3} =$

6. $\dfrac{1}{2} \div \dfrac{1}{4} =$

7. $21 \div \dfrac{5}{6} =$

8. $2\dfrac{5}{16} \div 13 =$

9. $57 \div 2\dfrac{5}{8} =$

10. $63 \div \dfrac{1}{63} =$

11. $9\dfrac{3}{16} \div 3\dfrac{9}{16} =$

12. $100 \div \dfrac{1}{2} =$

13. $1\dfrac{13}{16} \div 2\dfrac{1}{4} =$

14. $5\dfrac{4}{9} \div 5\dfrac{4}{9} =$

15. $\dfrac{7\dfrac{9}{16}}{7} =$

16. $\dfrac{4\dfrac{2}{5}}{2\dfrac{3}{4}} =$

17. $\dfrac{3\dfrac{5}{6}}{9\dfrac{1}{5}} =$

18. $\dfrac{\dfrac{123}{716}}{\dfrac{82}{182}} =$

19. How many 3 1/2″ pieces can be sheared from a thin piece of sheet metal 31″ long?

20. A piece of 1/2″ diameter copper tubing 105 1/2″ long is cut with a pipe cutter into six pieces of equal length. What is the length of each piece?

Use this information to answer questions 21 and 22. A bus, rented by E & P Mold Co., is to deliver a crew of workers to a job site 87 1/2 miles away. The bus departs from the shop at 6:30 a.m. and is expected to arrive at 9:00 a.m.

21. What average speed must the bus maintain to arrive on schedule?

22. If the bus averages 19 1/2 miles per gallon, how many gallons of fuel are needed for one trip to the site and back to the shop?

23. A sheet of metal weighs 18 7/16 lb. In a shearing operation, the sheet is cut into strips weighing 3/8 lb each. How many strips of metal are produced?

24. A steel plate 10′ 6 1/2″ long weighs 366 17/20 lb. How much does a 1″ length of the plate weigh?

25. A volume of 1 cubic foot contains about 7 1/2 gallons. How many cubic feet of oil will a cooling tank hold if it contains 1,776 1/2 gallons?

Name _____ **Date** _____ **Class** _____

26. Divide this T-shape into 13 equal parts. What length are the pieces?

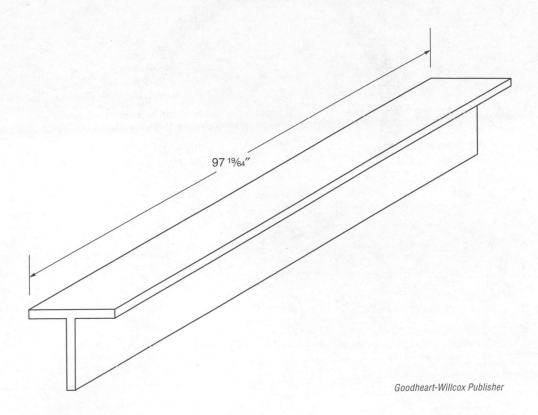

97 $^{19}/_{64}$"

Goodheart-Willcox Publisher

27. The following square bar of hot rolled steel is cut into four equal length pieces. Each saw cut wastes 3/32" of material. What length are the pieces?

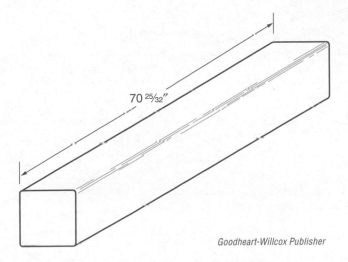

70 $^{25}/_{32}$"

Goodheart-Willcox Publisher

28. The cross section of a piece of extra strong pipe has the following dimensions. What is the wall thickness of the pipe?

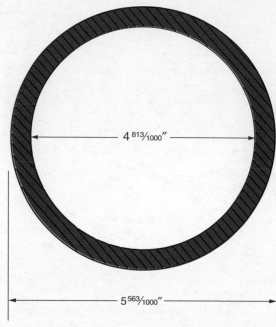

Goodheart-Willcox Publisher

29. How many strips 48″ long and 13 7/16″ wide can be sheared from this sheet of 15 gage steel?

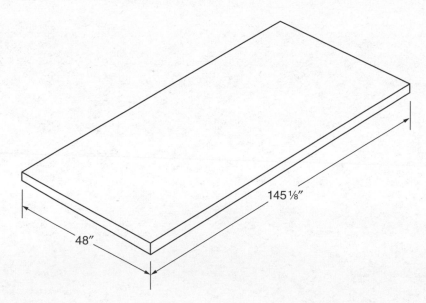

Goodheart-Willcox Publisher

Name _____   **Date** _____   **Class** _____

30. How many 7 3/4″ square pieces can be cut from this sheet?

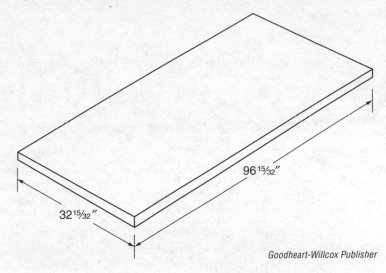

31. In the following diagram, fifteen more pieces of square tubing will be welded to the top side of this plate. Each tube will be spaced an equal distance from the next tube. Calculate the distance between any two consecutive tube pieces.

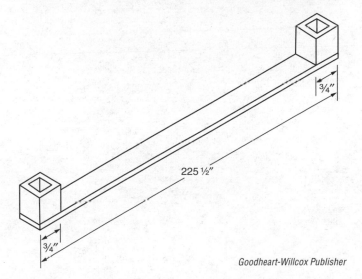

32. Three more rods of 1 1/4″ diameter and four more rods of 1 3/4″ diameter are to be welded to this plate. Center lines are to be equally spaced. What is the distance between centers?

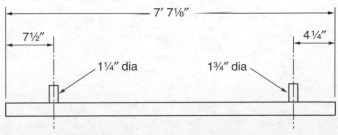

SECTION 3
Decimal Numbers

After studying this section, you will be able to:

- Give examples of decimal numbers.
- Explain how to convert decimal numbers and common fractions.
- Show how to round decimal numbers.
- Perform addition of decimal numbers.
- Perform subtraction of decimal numbers.
- Perform multiplication of decimal numbers.
- Perform division of decimal numbers.

Introduction to Decimal Numbers

Key Terms

decimal number   decimal point   rounding the decimal

Introduction

As you have seen so far, numbers can be classified in a variety of ways, such as whole numbers, common fractions, and mixed numbers. This unit introduces another classification of numbers called **decimal numbers**.

In Unit 1, it was noted that our numbering system is based on ten digits and that larger numbers are created by lining up these digits in a certain order. The decimal system uses this method of place position values to express numbers that are *less* than whole numbers. These are called *decimals*.

Decimal numbers include a **decimal point**. Digits to the left of the decimal point are whole numbers. Always include a zero to the left of the decimal point if there are no whole numbers in the units place.

Digits to the right of the decimal point are fractional numbers. Each position to the right of the decimal point has a value. Listed below are some of the names and place values.

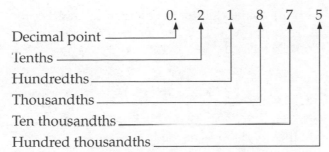

Goodheart-Willcox Publisher

There are two ways of expressing decimal numbers verbally:

- Pronounce each digit individually. For example, "14.63" is spoken as "fourteen point (or decimal) six three."

- Pronounce the decimal number as a whole number and add the name of the last place value. For example, "14.63" is spoken as "fourteen and sixty-three hundredths."

Converting Decimal Numbers to Common Fractions

A decimal number is changed to a common fraction by using the last place value as the denominator and the decimal number digits as the numerator.

$$0.3 = \frac{3}{10}$$

$$0.27 = \frac{27}{100}$$

$$0.173 = \frac{173}{1,000}$$

The task is simplified by the fact that denominators are always multiples of 10, such as 10, 100, 1,000, 10,000, etc. The actual value depends on the place position of the last digit. Once you have done the conversion, reduce the fraction if possible.

$$0.0625 = \frac{625}{10,000} = \frac{25}{400} = \frac{1}{16}$$

$$0.828125 = \frac{828,125}{1,000,000} = \frac{33,125}{40,000} = \frac{53}{64}$$

Converting Common Fractions to Decimal Numbers

A common fraction is changed to a decimal number by dividing the numerator by the denominator:

$$\frac{3}{4} = 4\overline{)3}$$

Place a decimal point to the right of the dividend. Place another decimal point directly above that decimal point. This is done so the decimal point in the answer (the quotient) is properly located:

$$\frac{3}{4} = 4\overline{)3.}$$

Now, add a zero to the right of the decimal point in the dividend and begin to divide. Add zeros to the dividend as needed:

$$\frac{3}{4} = \begin{array}{r} .75 \\ 4\overline{)3.00} \\ -2\,8 \\ \hline 20 \\ -20 \\ \hline 0r \end{array}$$

Rounding Decimals

The previous example is typical of most fractions. However, there are two situations you should watch for and take special action. First, some common fractions result in an unending decimal number. For example, 1/3 converts to the unending decimal number 0.33333. Second, some common fractions convert to decimal numbers that are quite long and much more accurate than the original common fraction. Therefore, it is common practice to reduce such numbers to a degree of accuracy that is adequate. This process is called *rounding the decimal*.

Before rounding the decimal, determine the degree of accuracy required. Normally, the problem description indicates the degree of accuracy required or to what place the decimal number should be extended. Also, the accuracy required is usually indicated as a tolerance on the shop drawings you will be using. The term *tolerance* is explained later in this text.

Eliminate all digits beyond the required degree of accuracy. If the first number you eliminate is 5 or more, increase the final number in your answer by 1. The following is an example:

1. Round 75.13846 to the nearest hundredth.

2. 75.13~~846~~

3. Since 8, the first number eliminated, is greater than 5, increase the 3 to a 4.

4. The answer is 75.14.

Work Space/Notes

Unit 11 Practice

Name _____ **Date** _____ **Class** _____

Write the following as decimal numbers.

1. One hundred decimal zero one

2. Ninety-five hundredths

3. Fourteen decimal zero zero one two five

4. Three thousand two hundred nineteen decimal one two five

5. Decimal seven zero seven

6. One thousand nine hundred seventeen ten thousandths

7. Decimal eight six six

8. Five decimal five one five six three

Convert the following decimal numbers to common fractions. Reduce to lowest terms.

9. 0.45

10. 0.28125

11. 0.1335

12. 0.6875

13. 0.3333

14. 0.78125

Convert the following to decimal numbers. Box your answers.

15. $\dfrac{1}{2}$

16. $100\dfrac{1}{100}$

17. $\dfrac{1}{32}$

18. $\dfrac{45}{64}$

19. $\dfrac{5}{8}$

20. $\dfrac{3}{1,000}$

Round the following decimals to the nearest tenth. Box your answers.

21. 0.8801

22. 0.1718

23. 0.4691

24. 0.5555

25. 0.0923

26. 0.0505

Round the following decimals to the nearest hundredth. Box your answers.

27. 0.9551

28. 0.02192

29. 0.0916

30. 0.0747

31. 0.9097

32. 0.9999

Round the following decimals to the nearest thousandth. Box your answers.

33. 0.3138

34. 0.09001

35. 0.0091

36. 0.4375

37. 0.4426

38. 0.0541

Convert the following common fractions to decimal numbers and round to the nearest hundredth.

39. $\dfrac{2}{3}$

40. $\dfrac{3}{4}$

41. $\dfrac{1,111}{10,000}$

42. $\dfrac{987}{1,000}$

43. $\dfrac{1,001}{10,000}$

44. $\dfrac{23}{43}$

Convert the following common fractions to decimal numbers and round to the nearest thousandth.

45. $\dfrac{15}{16}$

46. $\dfrac{3}{7}$

47. $\dfrac{1}{16}$

48. $\dfrac{21}{43}$

49. $\dfrac{7,666}{10,000}$

50. $\dfrac{75,196}{100,000}$

Convert the following common fractions to decimal numbers and round to the nearest ten thousandth.

51. $\dfrac{1}{64}$

52. $\dfrac{13}{15}$

53. $\dfrac{12,345}{100,000}$

54. $\dfrac{1}{9}$

55. $\dfrac{198}{205}$

56. $\dfrac{31}{32}$

Karimi Zetter/Shutterstock.com

UNIT 12

Addition and Subtraction of Decimal Numbers

Introduction

Addition and subtraction of decimal numbers are explained together in this unit. The two operations share important characteristics.

Method Used to Add Decimal Numbers

To add decimal numbers, write them in a column with the decimal points lined up, as follows:

$$
\begin{array}{r}
7.2 \\
19.01 \\
.6 \\
+\ 100.407 \\
\hline
\end{array}
$$

Be sure the decimal points are accurately lined up. You may find it helpful to add zeros so the far right side of the column is filled. Add all the digits starting from the right as if they are all whole numbers. Carry over any numbers necessary into the next column.

$$
\begin{array}{r}
11 \\
7.200 \\
19.010 \\
.600 \\
+\ 100.407 \\
\hline
127.217
\end{array}
$$

Method Used to Subtract Decimal Numbers

To subtract decimal numbers, you must also line up the decimal points. As in addition, you may add zeros to the far right side of numbers to fill the column and maintain decimal point alignment. Subtract each column starting from the right as if they are all whole numbers. When necessary, borrow from the column to the left.

$$
\begin{array}{r}
8\,1 \\
38.\cancel{9}47 \\
-\ 13.760 \\
\hline
25.187
\end{array}
$$

The operation is the same as that for whole numbers except for the decimal point. The values to the right of the decimal point are treated just the same as the values to the left of the decimal point. Refer to Unit 3 to review carrying to fill place values.

Work Space/Notes

Name _____ **Date** _____ **Class** _____

Perform the following equations. Show all your work. Be certain the columns line up.
Box your answers.

1. 0.3 + 9.2 + 3.9 =

2. 2.08 + 4.160 + 0.69 =

3. 0.354 + 354.009 + 9.035 =

4. 3.41
    2.45
    4.67
  + 5.26

5.   0.00532
     0.138
     4.322
   + 48.2

6.   0.109
     69.96
     10.01
    + 0.2

7. 16.601 + 0.195 + 4.749 + 0.945 + 200 =

8. 0.049 + 1,741.9 + 33.0 + 0.1 + 3.2 =

9. 0.6593 + 0.4978 + 100.0 + 3.1416 + 0.0101 =

10. 0.183 + 0.222 + 0.296 + 0.252 + 0.243 + 4.801 + 0.395 + 2.424 + 1.573 + 0.244 + 0.391 =

11.   0.9
    − 0.2

12.   19.51
    − 4.19

13.   60.0782
    − 42.38

14.   95.786
    − 88.999

15.   4.9
    − 0.807

16.   37.980
    − 37.421

17. 11.2 − 6.1356 =

18. 29.006 − 8.0005 =

19. 345.5842 − 0.095 =

20. A pipe has a wall thickness of 0.129″. The inside diameter is 3.716″. Find the outside diameter.

21. A bar of cast iron 18.125″ long has been tapered from a diameter of 2.74″ to a diameter of 0.983″. What is the difference in diameter between the two ends?

22. Kendan Stamping produces a metal oil pan for a cost of $12.97. To make a profit of $4.98, at what price should it be sold?

23. Two pieces of metal measuring 65.283″ and 23.014″ long are welded together using a root opening of 0.031″. What length is the final piece?

Use this information to answer questions 24 and 25. A salesperson estimated travel expenses for May at $675.00. At the end of May, the receipts were as follows—gasoline $120.59; meals $298.97; room $315.83; miscellaneous $37.16.

24. What were the total travel expenses?

25. By how much were the expenses greater or less than the estimated amount? Indicate a greater amount with a plus (+) and a lesser amount with a minus (–).

26. A steel pipe 17.074″ long has a 0.250″ cap welded to each end. The caps are then machined so 0.0625″ of material is removed from each cap. What is the final length of the weldment?

Name _____ **Date** _____ **Class** _____

Use the following diagram and this information to answer questions 27 through 35.
Each rod in this weldment is 0.6534" longer than the one before it. Rod A is 0.7596" long.

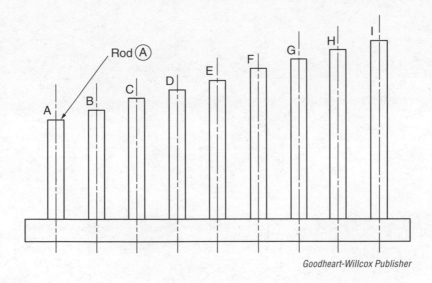

Goodheart-Willcox Publisher

27. Calculate the length of Rod Ⓑ.

28. Calculate the length of Rod Ⓒ.

29. Calculate the length of Rod Ⓓ.

30. Calculate the length of Rod Ⓔ.

31. Calculate the length of Rod (F).

32. Calculate the length of Rod (G).

33. Calculate the length of Rod (H).

34. Calculate the length of Rod (I).

35. Calculate the total length of rod used in this weldment.

Name _____   **Date** _____   **Class** _____

36. Calculate the length of (A) shown in the following illustration.

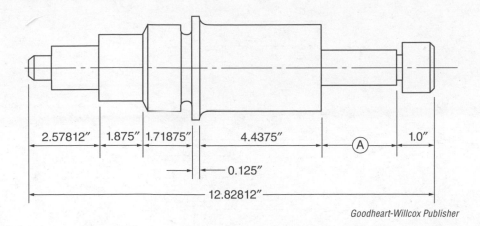

Use this information to answer questions 37 through 40. Review the following diagram. Grind 0.01562" from one side of each block and 0.02813" from the opposite side.

37. Calculate the finished thickness of block (A).

38. Calculate the finished thickness of block (B).

39. Calculate the finished thickness of block (C).

40. Calculate the combined finished thickness of all three blocks.

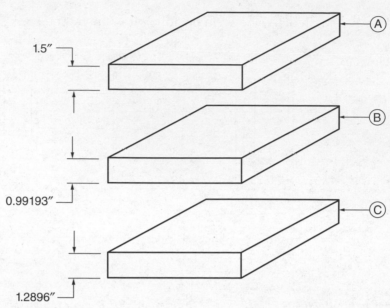

1.5″

0.99193″

1.2896″

Goodheart-Willcox Publisher

41. Calculate the combined length of the six contact surfaces shown in the following diagram.

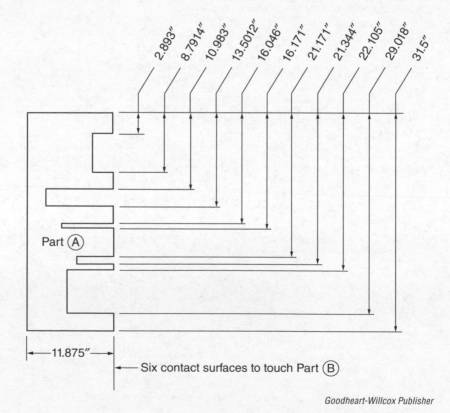

2.893″ 8.7914″ 10.983″ 13.5012″ 16.046″ 16.171″ 21.171″ 21.344″ 22.105″ 29.018″ 31.5″

Part Ⓐ

11.875″

Six contact surfaces to touch Part Ⓑ

Goodheart-Willcox Publisher

Name _____   **Date** _____   **Class** _____

42. What are the overall dimensions of the block shown in the following diagram after 0.892″ is machined from each surface?

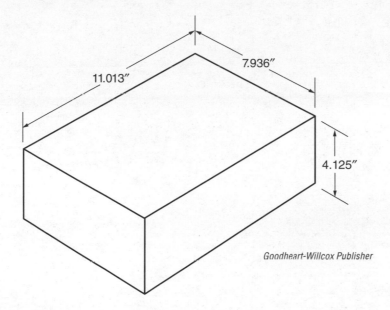

Goodheart-Willcox Publisher

Work Space/Notes

Karimi Zetter/Shutterstock.com

Multiplication of Decimal Numbers

Introduction

Multiplying decimal numbers is almost exactly the same as multiplying whole numbers. Refer to Unit 4 if you need to review the topic of multiplication.

Method Used to Multiply Decimal Numbers

When multiplying decimal numbers, the decimal points do not have to be lined up. Instead, line up the digits on the right side, as follows:

$$\begin{array}{r} 14.124 \\ \times \quad .31 \\ \hline \end{array}$$

Now, multiply in the same way you would for whole numbers. You can ignore the decimal point until you complete the multiplication process. Remember to use an X or 0 as a marker to align the columns of the preliminary product of individual multiplier digits.

$$\begin{array}{r} 14.124 \\ \times \quad .31 \\ \hline 14124 \\ 42372X \\ \hline 437844 \end{array}$$

When you have arrived at an answer, you will need to determine where the decimal point should be placed. To do this, count the total number of digits to the right of the decimal point in the two numbers being multiplied. In this example, the total is five. Now, count off five digits from the right in the answer and place the decimal point in front of (to the left of) the fifth digit. The answer here is 4.37844.

Occasions arise where you do not have enough digits in the answer to locate the decimal point. See the following example.

$$\begin{array}{r} .0234 \\ \times \quad .2 \\ \hline 0468 \end{array}$$

This problem contains a total of five digits to the right of the decimal point, but the answer produces only four digits. Whenever this occurs, add as many zeros as needed to the left of the answer. The answer here is 0.00468.

A different situation is illustrated by the next example.

$$
\begin{array}{r}
.875 \\
\times\ \ 2.4 \\
\hline
3500 \\
1750\text{X} \\
\hline
2.1000
\end{array}
$$

When the answer ends in zeros on the right side of the decimal point, the zeros at the end of the number can be eliminated. The answer here is 2.1.

This final example illustrates both adding zeros to the left and removing zeros from the right of the answer.

$$
\begin{array}{r}
.00684 \\
\times\ \ \ \ .25 \\
\hline
03420 \\
01368\text{X} \\
\hline
017100
\end{array}
$$

The answer is 0.00171.

Multiplication of Decimals by 10, 100, 1,000, Etc.

When a decimal number is multiplied by 10, 100, 1,000, 10,000, and so on, the answer can be quickly calculated. Move the decimal point to the *right* the same number of places as there are zeros in the multiplier. These examples will illustrate:

$$
\begin{aligned}
1.8 \times 10 &= 18.0 \\
1.8 \times 100 &= 180.0 \\
1.8 \times 1,000 &= 1,800.0 \\
.645 \times 100 &= 64.5 \\
2,600,000 \times 10,000 &= 26,000,000,000
\end{aligned}
$$

Name _____ **Date** _____ **Class** _____

Perform the following equations. Show all your work. Box your answers.

1. $\begin{array}{r} 0.89 \\ \times\, 0.34 \\ \hline \end{array}$

2. $\begin{array}{r} 0.409 \\ \times\, 9.04 \\ \hline \end{array}$

3. $\begin{array}{r} 0.572 \\ \times\, 123 \\ \hline \end{array}$

4. $\begin{array}{r} 7.008 \\ \times\, 0.006 \\ \hline \end{array}$

5. $\begin{array}{r} 10.0001 \\ \times\, 100.1 \\ \hline \end{array}$

6. $\begin{array}{r} 875 \\ \times\, 0.25 \\ \hline \end{array}$

7. $37 \times 0.18 =$

8. $18.125 \times 29 =$

9. $2{,}500 \times 0.4375 =$

10. $0.632 \times 22 =$

11. $0.52 \times 6.2 \times 9.3 =$

12. $0.66 \times 9.9 \times 12.0 =$

13. $0.02 \times 2.0 \times 0.002 =$

14. $0.5 \times 5 \times 0.05 =$

15. $569.59 \times 0.002 =$

16. $9.7825 \times 1.39 =$

17. $100{,}000 \times 1.95 =$

18. $0.0015 \times 100 =$

19. $2.423 \times 8.97 \times 3.0 =$

20. $10.01 \times 1{,}000 \times 1.0 =$

21. $11.1 \times 1.11 \times 0.111 =$

22. $22 \times 2.2 \times 0.22 =$

23. What is the weight of 2,906 aluminum castings if each casting weighs 23.47 pounds? Round your answer to the nearest 10 pounds.

24. One cubic inch of steel weighs 0.2835 pounds. What is the weight of a block of steel containing 35.48 in^3? Round your answer to the nearest hundredth of a pound.

25. A salesperson offered you the following deal on some shop supplies—$465.00 in cash or an $80.00 down payment with 12 monthly payments of $33.95 each. How much money would you save by paying cash?

26. What is the total cost to fill a storage tank with 5,385.5 gallons of diesel fuel at $4.87 per gallon? Express your answer in dollars and cents to the nearest cent.

27. A pipe support is welded in 3.75 minutes and uses 4.3125 ft^3 of acetylene gas. How many cubic feet of gas are used to weld 12 of the supports?

Name _____   **Date** _____   **Class** _____

28. An oxyacetylene cutting machine with eight cutting torches uses 2.975 ft³ of oxygen per torch to produce eight cover blanks. How many cubic feet of oxygen are required to produce 8,000 cover blanks?

29. A machining process at the Windsor Transmission Plant reduces a solid steel shaft from 103.301 pounds to 98.891 pounds. Every 60 minutes, 351.75 shafts are produced. At the end of 53.6 hours, what is the total weight of steel removed from the shafts?

Use this information to answer questions 30 through 32. Ten pieces of chain, each 6′ long, are required for a job. To clear out old inventory, the supply shop makes you a special offer—buy the remaining 64.5 feet of chain on the spindle at $3.74 per foot. Otherwise, you could buy only what you need at $3.91 per foot plus 60 cents ($0.60) per cut.

30. What is the total cost of the special offer?

31. What is the total cost of buying just what you need?

32. What is the cost saved by choosing the lower-priced option?

33. Review the following diagram. Seven tube assemblies are required to complete a job. What is the total length of condenser tube required?

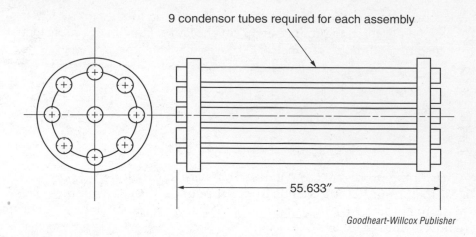

9 condensor tubes required for each assembly

55.633"

Goodheart-Willcox Publisher

34. Review the following diagram. Calculate the total weight of the weldments in Job #8861.

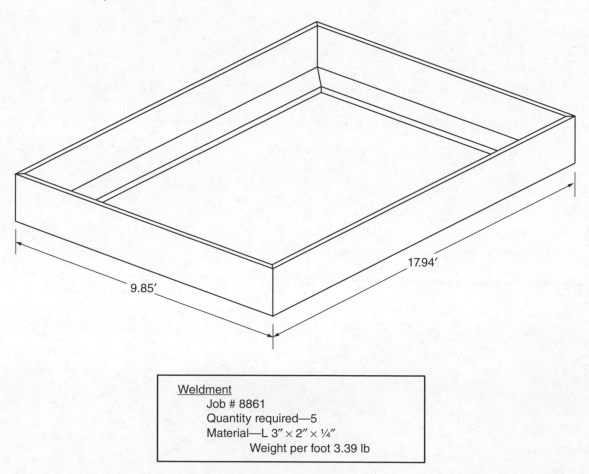

9.85'

17.94'

Weldment
Job # 8861
Quantity required—5
Material—L 3″ × 2″ × ¼″
Weight per foot 3.39 lb

Goodheart-Willcox Publisher

Name _____ **Date** _____ **Class** _____

35. Review the following diagram. Calculate the length of (A).

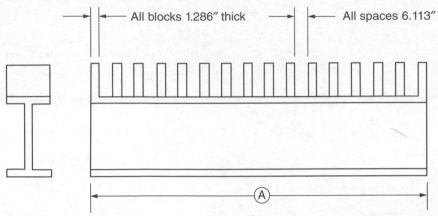

Goodheart-Willcox Publisher

36. Review the following diagram. Twenty of these braces are required. What is the total weight?

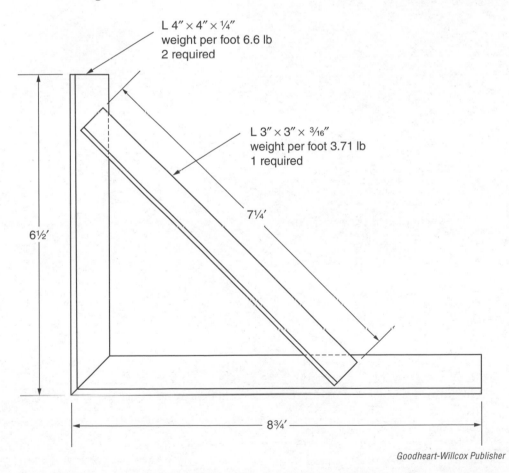

Goodheart-Willcox Publisher

37. Review the following diagram. The tank is made of 0.125″ thick steel. A 4′ wide sheet of steel weighs 20.412 pounds per foot. The base is made of 2″ square tubing with a wall thickness of 0.25″, which weighs 5.41 lb per foot. Water weighs 62.428 lb per cubic foot. The tank holds 26.56 ft³. Given this information, calculate the total weight of the base and the cooling tank, assuming that it is completely filled with water.

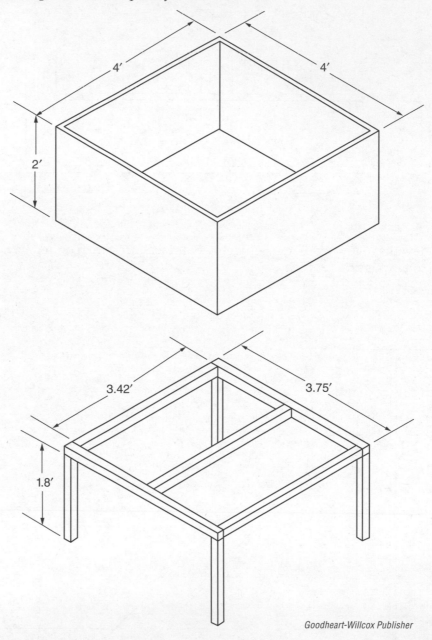

Goodheart-Willcox Publisher

38. Review the following order and the reference table provided. Should a truck designed for a maximum load of 10,000 pounds be used to deliver the following order in one trip?

20 pieces of 1 9/16″ round bar at 12′

500 pieces of 1/4″ round bar at 9′

70 pieces of 2″ square bar at 6′

5 pieces of 2 13/16″ round bar at 8′

Name _____   **Date** _____   **Class** _____

Square and Round Bars Weight and Area								
	Weight (lb per foot)		**Area (in²)**			**Weight (lb per foot)**		**Area (in²)**	
Size (in)	■	●	▨	◍	**Size (in)**	■	●	▨	◍
0					**3**	30.63	24.05	9.000	7.069
1/16	0.013	0.010	0.0039	0.0031	1/16	31.91	25.07	9.379	7.366
1/8	0.053	0.042	0.0156	0.0123	1/8	33.23	26.10	9.766	7.670
3/16	0.120	0.094	0.0352	0.0276	3/16	34.57	27.15	10.160	7.980
1/4	0.213	0.167	0.0625	0.0491	1/4	35.94	28.23	10.563	8.296
5/16	0.332	0.261	0.0977	0.0767	5/16	37.34	29.32	10.973	8.618
3/8	0.479	0.376	0.1406	0.1105	3/8	38.76	30.44	11.391	8.946
7/16	0.651	0.512	0.1914	0.1503	7/16	40.21	31.58	11.816	9.281
1/2	0.851	0.668	0.2500	0.1963	1/2	41.68	32.74	12.250	9.621
9/16	1.077	0.846	0.3164	0.2485	9/16	43.19	33.92	12.691	9.968
5/8	1.329	1.044	0.3906	0.3068	5/8	44.71	35.12	13.141	10.321
11/16	1.608	1.263	0.4727	0.3712	11/16	46.27	36.34	13.598	10.680
3/4	1.914	1.503	0.5625	0.4418	3/4	47.85	37.58	14.063	11.045
13/16	2.246	1.764	0.6602	0.5185	13/16	49.46	38.85	14.535	11.416
7/8	2.605	2.046	0.7656	0.6013	7/8	51.09	40.13	15.016	11.793
15/16	2.991	2.349	0.8789	0.6903	15/16	52.76	41.43	15.504	12.177
1	3.403	2.673	1.0000	0.7854	**4**	54.44	42.76	16.000	12.566
1/16	3.841	3.017	1.1289	0.8866	1/16	56.16	44.11	16.504	12.962
1/8	4.307	3.382	1.2656	0.9940	1/8	57.90	45.47	17.016	13.364
3/16	4.798	3.769	1.4102	1.1075	3/16	59.67	46.86	17.535	13.772
1/4	5.317	4.176	1.5625	1.2272	1/4	61.46	48.27	18.063	14.186
5/16	5.862	4.604	1.7227	1.3530	5/16	63.28	49.70	18.598	14.607
3/8	6.433	5.053	1.8906	1.4849	3/8	65.13	51.15	19.141	15.033
7/16	7.032	5.523	2.0664	1.6230	7/16	67.01	52.63	19.691	15.466
1/2	7.656	6.013	2.2500	1.7671	1/2	68.91	54.12	20.250	15.904
9/16	8.308	6.525	2.4414	1.9175	9/16	70.83	55.63	20.816	16.349
5/8	8.985	7.057	2.6406	2.0739	5/8	72.79	57.17	21.391	16.800
11/16	9.690	7.610	2.8477	2.2365	11/16	74.77	58.72	21.973	17.257
3/4	10.421	8.185	3.0625	2.4053	3/4	76.78	60.30	22.563	17.721
13/16	11.179	8.780	3.2852	2.5802	13/16	78.81	61.90	23.160	18.190
7/8	11.963	9.396	3.5156	2.7612	7/8	80.87	63.51	23.766	18.665
15/16	12.774	10.032	3.7539	2.9483	15/16	82.96	65.15	24.379	19.147
2	13.611	10.690	4.0000	3.1416	**5**	85.07	66.81	25.000	19.635
1/16	14.475	11.369	4.2539	3.3410	1/16	87.21	68.49	25.629	20.129
1/8	15.366	12.068	4.5156	3.5466	1/8	89.38	70.20	26.266	20.629
3/16	16.283	12.788	4.7852	3.7583	3/16	91.57	71.92	26.910	21.135
1/4	17.227	13.530	5.0625	3.9761	1/4	93.79	73.66	27.563	21.648
5/16	18.197	14.292	5.3477	4.2000	5/16	96.04	75.43	28.223	22.166
3/8	19.194	15.075	5.6406	4.4301	3/8	98.31	77.21	28.891	22.691
7/16	20.217	15.879	5.9414	4.6664	7/16	100.61	71.92	29.566	23.221
1/2	21.267	16.703	6.2500	4.9087	1/2	102.93	80.84	30.250	23.758
9/16	22.344	17.549	6.5664	5.1572	9/16	105.29	82.69	30.941	24.301
5/8	23.447	18.415	6.8906	5.4119	5/8	107.67	84.56	31.641	24.850
11/16	24.577	19.303	7.2227	5.6727	11/16	110.07	86.45	32.348	25.406
3/4	25.734	20.211	7.5625	5.9396	3/4	112.50	88.36	33.063	25.967
13/16	26.917	21.140	7.9102	6.2126	13/16	114.96	90.29	33.785	26.535
7/8	28.126	22.090	8.2656	6.4918	7/8	117.45	92.24	34.516	27.109
15/16	29.362	23.061	8.6289	6.7771	15/16	119.96	94.22	35.254	27.688
3	30.625	24.053	9.0000	7.0686	**6**	122.50	96.21	36.000	28.274

American Institute of Steel Construction

39. Review the following parts list. What is the total weight of Job #8853?

Parts List Job—8853 Date—August 29		
Quantity	Part Description	Weight
37	3/4″ STD pipe at 6.25′	1.13 lb per foot
148	1″ round bar at 0.5685′	2.67 lb per foot
4	M shape at 10.5′	18.9 lb per foot
7	L 5″ × 3″ × 1/4″ at 7.0625′	6.6 lb per foot
8	3″ STD channel at 8.125′	4.1 lb per foot

Division of Decimal Numbers

Introduction

Dividing decimal numbers is almost exactly the same as dividing whole numbers. Refer to Unit 5 to review the topic of division.

Method Used to Divide Decimal Numbers

When dividing decimal numbers, arrange the dividend and divisor as you would for whole number division. However, your first concern is the decimal point. If there is a decimal point in the divisor, move it to the right of the far right digit. You want to change the divisor to a whole number. Then move the decimal point in the dividend the same number of places to the right. If necessary, add zeros to the right of the decimal point. Here is an example.

Divide 186.942 by 17.34 to the nearest thousandth. Begin with the division equation setup.

$$17.34\overline{)186.942}$$

Move the decimal point in the divisor (17.34) to the far right. Then move the decimal point in the dividend (186.942) the same number of places to the right.

$$1734.\overline{)18694.2}$$

Shifting the decimal point changes the divisor to a whole number and greatly simplifies the task of division. Next, place a decimal point directly above the decimal point in the dividend. This is done so the decimal point in the answer is properly located.

$$1734.\overline{)18\overset{.}{6}94.2}$$

Now, proceed with the division. Remember to calculate to the ten thousandth so the answer can be rounded to the nearest thousandth.

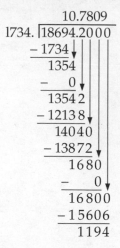

The quotient 10.7809 is then rounded to the nearest thousandth. The 9 in the ten thousandth place rounds the 0 in the thousandth place up to 1. The answer is 10.781.

As shown in the example, the dividend may not have enough zeros to accommodate the relocated decimal point. If this occurs, add enough zeros so you can locate the decimal point properly. See the following example.

$$9.345 \overline{)27.25}$$

The decimal point is relocated like this:

$$9345. \overline{)27250.}$$

Division of Decimals by 10, 100, 1,000, Etc.

When a decimal number is divided by 10, 100, 1,000, 10,000, and so on, the answer can be quickly calculated. Move the decimal point to the *left* the same number of places as there are zeros in the divisor. These examples will illustrate:

$$185 \div 10 = 18.5$$

$$185 \div 100 = 1.85$$

$$185 \div 1,000 = .185$$

$$968 \div 10,000 = .0968$$

$$7,500,000 \div 1,000,000 = 7.5$$

Name _____ **Date** _____ **Class** _____

Divide the following equations. Show all your work. Box your answers.

1. $875 \div 0.25 =$

2. $1.17 \div 0.003 =$

3. $3.94 \div 20 =$

4. $10.001 \div 0.01 =$

5. $0.072 \div 0.009 =$

6. $8.6752 \div 1{,}000 =$

Divide the following equations to the nearest hundredth. Show all your work. Box your answers.

7. $\dfrac{0.1217}{1.72} =$

8. $\dfrac{7.85}{2.16} =$

9. $\dfrac{100.449}{100} =$

10. $\dfrac{96{,}603}{100{,}000} =$

11. $\dfrac{6.3}{0.046} =$

12. $\dfrac{72}{0.71} =$

Divide the following equations to the nearest thousandth. Show all your work. Box your answers.

13. $16.5 \div 5.5 =$

14. $48 \div 10.1 =$

15. $17.1298 \div 1{,}000 =$

16. $0.96 \div 6.9 =$

17. $6.9 \div 9.6 =$

18. $2{,}009.15 \div 100 =$

19. A gear pump can move 113.4 gallons of water per hour. How many hours would it take to empty a tank containing 1,822 gallons? Express your answer to the nearest hour.

20. One carton containing nine one-gallon cans of cold galvanizing compound costs $137.25. How many one-gallon cans are you able to buy for $4,422.50?

21. A large container of stainless steel ball bearings weighs 212.55 pounds. Each ball bearing weighs 0.2834 pounds. How many ball bearings are in the container?

22. A surface grinder removes 0.013" of steel with each pass. How many passes are required to reduce a piece of steel from 2.095" to 1.718"?

Use this information to answer the two questions that follow. A special assignment includes producing precision cut blocks of steel measuring 0.869" thick. These blocks are going to be placed into an opening that is 3.652".

23. How many precision-cut blocks will fit in the opening?

24. How much space, if any, will remain in the opening after the blocks have been inserted?

25. A company offered the following deal on a standard item they produce. The first 150 items a customer ordered would cost $5,227.50. Each item after that would cost the customer $30.50. If the company received a $19,654.00 order, how many items were in the order?

Name _____   **Date** _____   **Class** _____

26. Refer to the following diagram. Divide this channel into 11 equal pieces.
    What is the length of each piece?

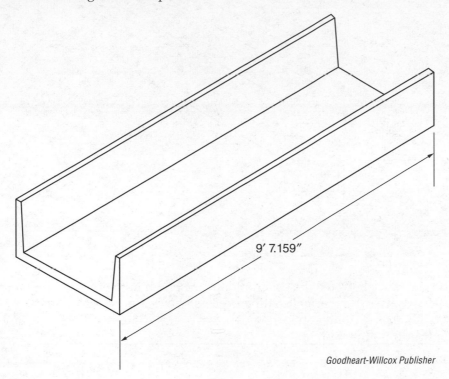

9′ 7.159″

Goodheart-Willcox Publisher

27. How many 1.857″ squares can be cut from this sheet of steel?

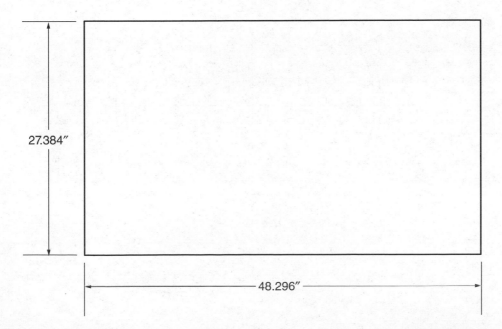

27.384″

48.296″

Goodheart-Willcox Publisher

28. This square bar weighs 156.686 pounds. What is the weight of a 1″ piece?

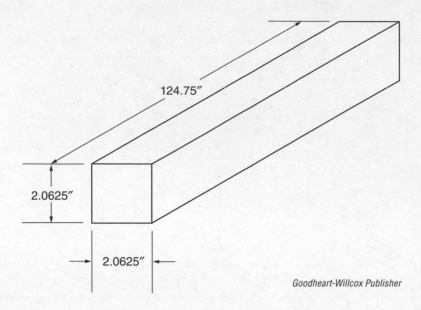

124.75″

2.0625″

2.0625″

Goodheart-Willcox Publisher

29. The clearance between the collar and drum shown in the following diagram is 0.0573″. Using a thermal coating process, the clearance is to be reduced to 0.0087″. Each coating application deposits 0.0009″ of material. How many applications are required?

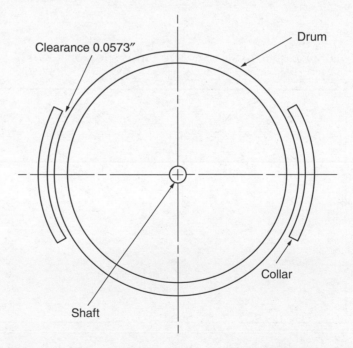

Clearance 0.0573″

Drum

Collar

Shaft

Goodheart-Willcox Publisher

Name _____   **Date** _____   **Class** _____

30. How many of the following drums would there be in a stack 73.0232″ high?

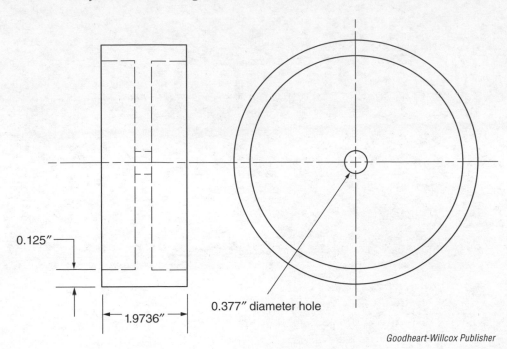

0.125″

1.9736″

0.377″ diameter hole

Goodheart-Willcox Publisher

31. What is the distance between the evenly spaced vertical blocks in the following diagram?

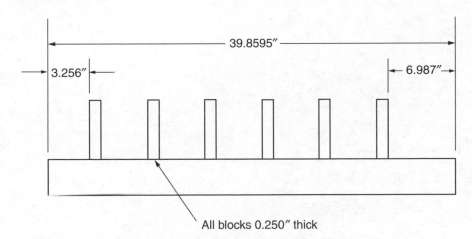

39.8595″

3.256″

6.987″

All blocks 0.250″ thick

Goodheart-Willcox Publisher

SECTION 4
Measurement

Section Objectives

After studying this section, you will be able to:

- Define linear measure, angular measure, and circular measure.
- Convert linear measurements between units.
- Explain and give examples of various forms of tolerance.
- Perform angular measurement.
- Perform four basic operations for angular calculations.
- Define basic four-sided shapes.
- Calculate perimeter and area for four-sided shapes.
- Convert square units of measure.
- Define basic triangular shapes.
- Calculate perimeter and area for triangular shapes.
- Define parts of circular shapes.
- Calculate circumference and area of circular shapes.

UNIT 15

Linear Measure

Key Terms

angular measure

circular measure

linear measure

millimeter (mm)

SI (metric) system

tolerance

US customary system

Introduction

A welder's job includes taking and reading measurements. Generally, as a welder, you will work with three types of measurements:

- **Linear measure** refers to measuring the straight line distance between two points.
- **Angular measure** refers to measuring the angle formed by two intersecting lines.
- **Circular measure** refers to measuring curved lines.

Two linear measuring systems are presently in use by welders—the US customary system and the SI (metric) system. You should become very familiar with both systems.

US Customary System

The **US customary system** is the measuring system commonly used in the United States. The basic unit of length is the inch. Twelve inches make up one foot, and 36 inches equal one yard. In welding, you will be concerned only with feet and inches. For example, a standard notation is 6'-11". This notation lists feet before inches and connects them with a hyphen (-).

How to Read a Measuring Tape (US Customary)

Welders typically use a 6' measuring tape and a 6" steel ruler. The measuring tape is divided into feet and inches, and the ruler is divided into inches. Each inch is identified by a number. The fractional parts of each inch are identified by short vertical lines of varying length. The length indicates what fractional part of the inch it represents. Here is how it works.

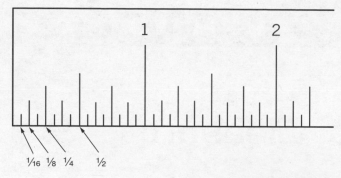

On a typical ruler, the smallest fractional part of an inch indicated is a sixteenth. However, rulers are available with inches divided into thirty-seconds or sixty-fourths.

If you not sure what the smallest division of a ruler is, start reading the marks within the inch division. The longest line is at the half point. The next longest line is at the quarter point, the next is at the 1/8, the next at the 1/16. If that is the smallest division, your ruler is divided into sixteenths. With practice, you will be able to read these markings quickly.

Since linear measure is so important to your work, take your time when measuring. If you cannot keep the numbers in your memory, write them down. Always measure twice!

SI Metric System

You are taking your training at a time when the welding industry, along with the rest of industrial America, is changing from the US customary system to the *SI (metric) system*. A full explanation of metrics as it relates to your work is provided in Unit 24. A short explanation of metric linear measurement follows.

The Millimeter

The smallest unit of length you will encounter in the shop or in the field is the *millimeter*. In fact, it is probably the only metric unit you will use because almost all metric blueprints are dimensioned only in millimeters. It is the basic unit in metric dimensioning, just as the inch is the basic unit in the US customary system. The notation used for millimeters is *mm*. Note that metric dimensions are always in decimal form.

As you can see in the following illustration, a millimeter is a very small unit of measurement. Here are two ways to visualize a millimeter.

- A millimeter is approximately the thickness of a line made with a dull pencil.
- A dime is about one millimeter thick.

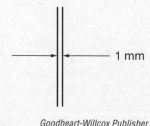

Since millimeters are small, they accumulate very quickly. For example, there are 25.4 millimeters in one inch. If you are 5′-10″ tall, then you are 1778 mm tall. A 10′ bar is 3048 mm long. Eventually, through practice, you will develop a "feel" for metric lengths. If you do not own a metric scale, you should acquire one now. Note that in the metric system, a comma is not placed between sets of three digits.

How to Read a Metric Measuring Tape

The metric scale is much simpler than an inch/foot scale. The shortest lines on the scale represent millimeters. Medium-length lines indicate 5 mm. The longest lines indicate 10 mm and are identified by a number. An example follows.

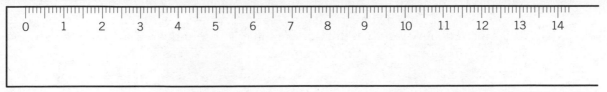

Goodheart-Willcox Publisher

The 1 indicates 10 mm, the 2 indicates 20 mm, and so on. As you will learn in Unit 24, 10 mm is equal to 1 cm. However, when reading the metric scale, all dimensions are referred to in millimeters. The 9 is read as 90 mm, the 10 is read as 100 mm, and so on. In the following example, the long dimension is identified as 28 mm, not as 2 cm and 8 mm.

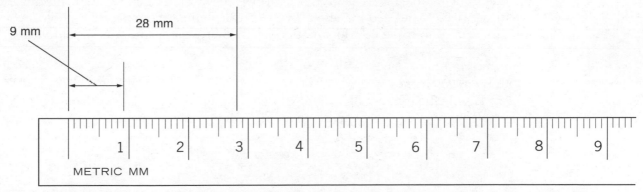

Goodheart-Willcox Publisher

Some metric scales appear as shown in the following example. The scale here is self-explanatory.

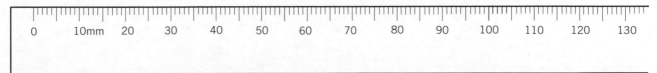

Goodheart-Willcox Publisher

Converting Linear Measurements

Your shop work will require you to change back and forth between inches, feet, and millimeters. Often, these conversions result in lengthy decimals, so it is common practice to round off conversions. Also, you will sometimes have to convert decimal dimensions to fractional dimensions (eighths, sixteenths, etc.).

Converting Feet to Inches

To convert from feet to inches, multiply the number of feet by twelve.

$$7' = 7 \times 12 = 84''$$

$$27.35' = 27.35 \times 12 = 328.2''$$

$$4\frac{19}{64}' = \frac{275}{64} \times 12 = \frac{3{,}300}{64} = 51\frac{36}{64} = 51\frac{9}{16}''$$

Converting Inches to Feet

To convert inches to feet, divide the number of inches by twelve. The answer is in its most useful form when expressed in feet and inches rather than only in feet. Note that the remainder of each division problem is the number of inches to be written after the number of feet in the notation.

$$109'' = 109 \div 12 = 9'\text{-}1''$$

$$219\frac{1}{4}'' = 219\frac{1}{4} \div 12 = 18'\text{-}3\frac{1}{4}''$$

$$28.7'' = 28.7 \div 12 = 2'\text{-}4.7''$$

As shown in the last two examples, the partial inch values did not need to be included in the calculation. This would have complicated the equation. These fractional inch values were easily brought over from the original value. Only whole inches need to be used in these conversion calculations, because only whole inches can be used to make up a foot measurement. Study these additional examples.

Example 1

Convert 159 inches to feet.

Divide the number of inches by twelve. Perform the division only as far as the whole number. Do not continue to the right of the decimal point.

$$
\begin{array}{r}
13 \\
12\,\overline{)159} \\
-12 \\
\hline
39 \\
-36 \\
\hline
3
\end{array}
$$

The answer is 13 3/12' or 13'-3".

Example 2

Convert 2,413 3/16 inches to feet.

First, perform the division only on the whole number.

$$
\begin{array}{r}
201 \\
12\,\overline{)2413} \\
-24 \\
\hline
013 \\
-12 \\
\hline
1
\end{array}
$$

The result of this division is 201 1/12'. Express this as 201'-1". Now, add the additional 3/16" to the answer. The final answer is 201'-1 3/16".

Example 3
Convert 899.17 inches to feet.
First, perform the division only on the whole number.

$$\begin{array}{r} 74 \\ 12\overline{\smash)899} \\ \underline{-84} \\ 59 \\ \underline{-48} \\ 11 \end{array}$$

The result of this division is 74 11/12', which is 74'-11". Next, add the 0.17" to the answer. The final answer is 74'-11.17".

Adding Feet and Inches

To add feet and inches, first line up the dimensions to be added in columns.

Example 1

$$\begin{array}{r} 7'\text{-}8'' \\ 32'\text{-}2'' \\ 111'\text{-}11'' \\ \underline{+\ 0'\text{-}9''} \\ 150'\text{-}30'' \end{array}$$

Add the numbers in the inches column. If the inches total 12 or more, convert the inches to feet.

$$30'' = 2'\text{-}6''$$

Then, add the numbers in the feet column. Finally, add the total of the feet column plus the feet and inches calculated from the inches column.

$$150'\text{-}0'' + 2'\text{-}6'' = 152'\text{-}6''$$

Example 2
If the inches include fractions, convert the fractions to a common denominator and then add.

$$\begin{array}{rcl} 3'\text{-}4\ 3/8'' & = & 3'\text{-}4\ 6/16'' \\ 1'\text{-}7\ 1/2'' & = & 1'\text{-}7\ 8/16'' \\ \underline{+\ 0'\text{-}9\ 3/16''} & = & \underline{+\ 0'\text{-}9\ 3/16''} \\ 4'\text{-}20\ 17/16'' & = & 5'\text{-}9\ 1/16'' \end{array}$$

Subtracting Feet and Inches

To subtract feet and inches, line up the dimensions to be subtracted in columns. First subtract the inches, and then subtract the feet.

Example 1

$$\begin{array}{r} 19'\text{-}11'' \\ \underline{-\ 1'\text{-}7''} \\ 18'\text{-}4'' \end{array}$$

Example 2

$$19'\text{-} 7''$$
$$\underline{-\ 1'\text{-}11''}$$

In this example, borrow 1′ (12″) from the 19′ and express the problem as shown:

$$18'\text{-}19''$$
$$\underline{-\ 1'\text{-}11''}$$
$$17'\text{-}8''$$

Example 3

If the inches include fractions, convert the fractions to a common denominator and then subtract.

$$22'\text{-}4\ 1/8'' \quad = \quad 21'\text{-}16\ 1/8'' \quad = \quad 21'\text{-}16\ 8/64''$$
$$\underline{-\ 5'\text{-}7\ 3/64''} \quad = \quad \underline{-\ 5'\text{-}7\ 3/64''} \quad = \quad \underline{-\ 5'\text{-}7\ 3/64''}$$
$$16'\text{-}9\ 5/64''$$

Multiplying Feet and Inches

Multiplication of feet and inches is generally easier if the dimensions are converted to inches. The final answer is then converted back to feet and inches. The following examples use multiplication to calculate length and area.

Length

In this example, the total length of seven spaces marked off along a factory floor is calculated. Each space is 10′-6 1/2″ in length. To calculate the total length, 10′-6 1/2″ is multiplied by seven.

First, convert feet and inches to inches:

$$10'\text{-}6\ 1/2'' = 126\ 1/2''$$

Then multiply:

$$126\frac{1}{2}'' \times 7 = \frac{253}{2}'' \times \frac{7}{1} = \frac{1{,}771}{2}''$$

Convert inches back to feet and inches:

$$\frac{1{,}771}{2}'' = 885\frac{1}{2}'' = 73'\text{-}9\frac{1}{2}''$$

Area

Area can be calculated in square feet or square inches. In this example, a length of 4′-6″ is multiplied by a width of 1′-9″.

Area calculated in square feet:

$$4'\text{-}6'' \times 1'\text{-}9''$$
$$4.5' \times 1.75' = 7.875\ \text{ft}^2$$

Area calculated in square inches:

$$4'\text{-}6'' \times 1'\text{-}9''$$
$$54'' \times 21'' = 1{,}134\ \text{in}^2$$

Dividing Feet and Inches

Division is generally easier if the dimensions are expressed in inches. The final answer is then converted back to feet and inches.

Example 1

$$13\text{'-}7\ 1/2'' \div 4 = 163\ 1/2 \div 4 = \frac{327}{2} \times \frac{1}{4} = \frac{327}{8} = 40\ 7/8'' = 3\text{'-}4\ 7/8''$$

Example 2

In many cases, the division is performed more easily by converting to decimal inches.

$$11\text{'-}4\ 3/16'' \div 6 = 136.1875'' \div 6 = 22.6979''$$

For normally acceptable shop accuracy, the answer can be rounded to the nearest 16th.

$$0.69 = \frac{69}{100} = \frac{x}{16} = 69 \times \frac{16}{100} = \frac{1{,}104}{100} = 11\frac{4}{100} = \frac{11}{16}$$

The answer is $22\frac{11}{16}''$.

Converting Inches to Millimeters

To convert from inches to millimeters, multiply the number of inches by 25.4. A good rule for rounding the answer is to express it with one *less* decimal place than the question.

$$16'' = 16 \times 25.4 = 406.4 = 406 \text{ mm}$$

$$93.375'' = 93.375 \times 25.4 = 2371.725 = 2371.73 \text{ mm}$$

$$\frac{1}{16}'' = \frac{1}{16} \times 25.4 = 1.5875 = 1.588 \text{ mm}$$

Converting Millimeters to Inches

To convert from millimeters to inches, divide the number of millimeters by 25.4. A good rule for rounding the answer in inches is to express it with one *more* decimal place than the question.

$$1789 \text{ mm} = 1789 \div 25.4 = 70.4''$$

$$9.32 \text{ mm} = 9.32 \div 25.4 = .3669 = .367''$$

$$360.5 \text{ mm} = 360.5 \div 25.4 = 14.19''$$

Summary of Conversion Factors		
From	**To**	**Method**
Feet	Inches	Multiply by 12
Inches	Feet	Divide by 12
Inches	Millimeters	Multiply by 25.4
Millimeters	Inches	Divide by 25.4

Goodheart-Willcox Publisher

Converting Decimals to Fractions

In Unit 11, you learned to convert decimals to fractions. A new situation arises when doing such a conversion for the purpose of linear measure. Here, you will have to convert the decimal dimension to one of the fractional dimensions normally found on your steel rule, such as eighths, sixteenths, thirty-seconds, or sixty-fourths. A typical example would be to convert 10.21″ to a fractional dimension to the nearest sixty-fourth.

1. Dealing with the decimal part only, 0.21 converts to 21/100.
2. Since 21/100 will not reduce exactly to a fraction with a denominator of 64, it will have to be reduced to the nearest sixty-fourth.
3. Write an equation as shown:

$$\frac{21}{100} = \frac{x}{64}$$

Since you want the final fraction to be in sixty-fourths, you know the denominator will be 64. However, you do not yet know what the numerator will be, so for now it is identified as x.

4. The method for finding x can be expressed in one phrase—*cross multiply and divide*.

1. $\dfrac{21}{100} \searrow \dfrac{x}{64}$

2. $21 \times 64 = 1{,}344$

3. $1{,}344 \div 100 = 13\dfrac{11}{25}$

Remember, you are trying to find the number that will replace the x in the fraction $x/64$. That number as calculated so far would be 13 11/25. But, instead of using 13 11/25, round to the nearest whole number, which is 13, since 11 is closer to 0 than to 25. Thirteen, then, is the numerator.

5. The answer is 10 13/64″ (to the nearest sixty-fourth).

Tolerance

Taking accurate measurements is important to any tradesperson. But there is a limit to how accurately you can measure. This limit is determined by two main factors—the accuracy of the measuring tool and the accuracy of the surface of the material being measured. Because of this, dimensions on blueprints normally indicate a certain acceptable range that you are allowed to work within. This range is called the *tolerance*, and it is usually indicated as shown in the following examples.

Tolerance +1/8″

This indicates an object may exceed the given dimension by 1/8″. Given a tolerance of +1/8″, an object dimensioned 48 3/8″ long would have a maximum acceptable dimension of 48 1/2″ and a minimum of 48 3/8″.

Tolerance –0.0625″

This indicates an object may be 0.0625″ less than the given dimension. Given a tolerance of –0.0625″, an object dimensioned 99.6″ long would have a maximum acceptable dimension of 99.6″ and a minimum of 99.5375″.

Tolerance ±3 mm

This indicates an object may be 3 mm greater or less than the given dimension. In this case, you have a range of 6 mm to work within. Given a tolerance of ±3 mm, an object dimensioned 211 mm long would have a maximum acceptable dimension of 214 mm and a minimum of 208 mm.

Work Space/Notes

Name _____ Date _____ Class _____

Measure the following lines as accurately as possible, in millimeters and inches.
Box your answers.

1. ———————————————————

2. —

3. ———————————

4. ——————————————————————

5. ———

6. ———————————

Measure the following as accurately as possible in millimeters and inches.

7. Your height: _____ mm, _____ in.

8. The height of a coffee cup: _____ mm, _____ in.

9. The length of a welding rod: _____ mm, _____ in.

10. The height of a workbench or desk: _____ mm, _____ in.

11. The span of your hand from smallest finger to thumb: _____ mm, _____ in.

12. The width of a doorway: _____ mm, _____, in.

Convert the following values from feet to inches. Show all your work.
Box your answers.

13. 35 1/4′   14. 123′

15. 45.375′   16. 17 5/23′

17. 503.5′   18. 4 1/8′

19. 5,280′   20. 65.1′

21. 17/64′   22. 29.47′

23. 12 1/12′   24. 15/32′

Convert the following inch values to feet (and inches, if necessary). Show all your work. Box your answers.

25. 108″   26. 387″

27. 4,801″   28. 6,000″

29. 215 3/4″   30. 12″

31. 787.96″   32. 17 1/2″

33. 4 1/8″   34. 38 1/4″

35. 17/64″   36. 0.29″

Convert the following inch values to millimeters. Round to the nearest tenth of a millimeter. Show all your work. Box your answers.

37. 12″   38. 36″

39. 10″   40. 1″

41. 1/64″   42. 69″

43. 5/8″   44. 1,000″

45. 63 3/16″   46. 59″

47. 8 1/4″   48. 480″

Convert the following millimeters to the decimal equivalent in inches. Round to the nearest thousandth of an inch. Show all your work. Box your answers.

49. 1000 mm   50. 25.4 mm

51. 1.0 mm   52. 279.4 mm

53. 65.5 mm   54. 12 mm

55. 11850 mm   56. 35600 mm

57. 195 mm   58. 862.3 mm

59. 5280 mm   60. 100 mm

Convert the following decimal inches to fractional dimensions. Calculate to the nearest eighth of an inch. Show all your work. Box your answers.

61. 15.8″   62. 0.982″

63. 48.012″   64. 80.125″

65. 19.375″   66. 0.9126″

Name _____ **Date** _____ **Class** _____

Convert the following values to the nearest sixteenth of an inch. Show all your work.
Box your answers.

67. 19.85″   68. 32.032″

69. 0.997″   70. 855.5″

71. 21.0625″   72. 1,010.7″

Convert to the nearest thirty-second of an inch. Show all your work. Box your answers.

73. 11.65″   74. 58.022″

75. 730.18″   76. 0.28125″

77. 0.979″   78. 3,602.79″

Convert to the nearest sixty-fourth of an inch. Show all your work. Box your answers.

79. 68.033″   80. 0.908″

81. 12.55″   82. 0.578125″

83. 632.41″   84. 7,802.77″

85. Lunar–Bray Mfg. ordered five beams (S 3 × 5.7 lb/ft) of the following lengths:
2′-6 1/2″, 3′-3/4″, 5′-2″, 2′-5″, 1′-6 5/8″. What is the total length of the order?

86. A rush order for three metal straps was received late Friday afternoon. They
measured 4′-1 1/4″, 4′-7 5/8″, 7′-1/2″ and were to be sheared from a piece
measuring 16′-0″. How much of the piece remained after filling the order?

87. A tool crib measures 15′-3″ by 23′-9″. What is the area of the room in square
feet? What is the area of the room in square inches?

88. What is the area in square inches of a plate measuring 1′-9 1/4″ × 1′-6 1/2″?

89. Nine pieces are to be sheared from a length of steel rod. Each piece is 1′-5 3/8″
long. What total length would the nine pieces use? Express your answer in
feet and inches.

90. A 12′-6″ strip of aluminum 3″ wide and 0.125″ thick is to be evenly divided
into 11 pieces. What is the length of each piece? Express your answer to the
nearest 1/32″.

Perform calculations and fill in the blanks in the following table. Determine the maximum and minimum dimensions allowed in the following lengths.

	Dimension	Tolerance	Maximum	Minimum
91.	249 mm	− 2 mm		
92.	17 1/32″	+ 1/16″		
93.	1.340″	± 0.005″		
94.	8′-11 3/4″	+ 1/2″		
95.	14.5 mm	± 0.5 mm		
96.	18 9/16″	+ 1/32″		
97.	19.9 mm	± 0.2 mm		
98.	8′-0″	− 1/16″		

Goodheart-Willcox Publisher

Review the following diagram. Convert its dimensions to fractional dimensions (to the nearest one-hundredth of an inch). Then, using a tolerance of ± 3/100″, calculate the maximum and minimum allowable dimensions. Record your answers in the table provided.

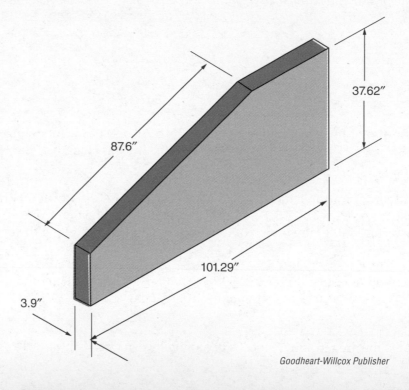

Goodheart-Willcox Publisher

	Decimal Dimension	Fractional Dimension	Tolerance	Maximum	Minimum
99.	3.9″		± 3/100″		
100.	37.62″		± 3/100″		
101.	87.6″		± 3/100″		
102.	101.29″		± 3/100″		

Goodheart-Willcox Publisher

Name _____   **Date** _____   **Class** _____

Review the following diagram. The tolerance for the hole in this weldment is +1/8".
All other dimensions have a tolerance of ± 1/16". Calculate the maximum and minimum
allowable dimensions and record your answers in the table provided.

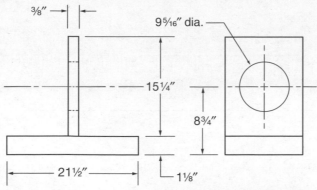

Goodheart-Willcox Publisher

	Dimension	Tolerance	Maximum	Minimum
103.	3/8″	± 1/16″		
104.	1 1/8″	± 1/16″		
105.	8 3/4″	± 1/16″		
106.	9 5/16″	+ 1/8″		
107.	15 1/4″	± 1/16″		
108.	21 1/2″	+ 1/16″		

Goodheart-Willcox Publisher

109. Measure and record the following distances in inches.

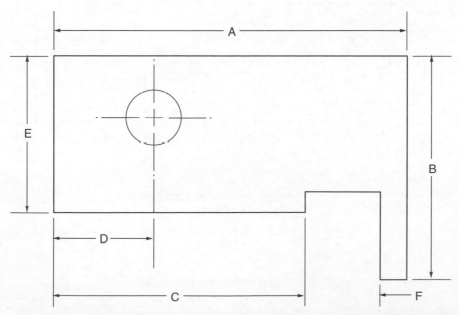

Goodheart-Willcox Publisher

110. Read and record the following distances in inches.

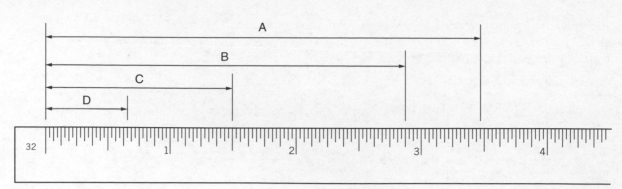

111. Using a straightedge, draw precise, sharp lines of the following lengths.

A. 6 3/16″   D. 5 6/16″

B. 7/8″   E. 2 1/4″

C. 7/16″   F. 2 1/2″

112. Measure and record the following distances in millimeters.

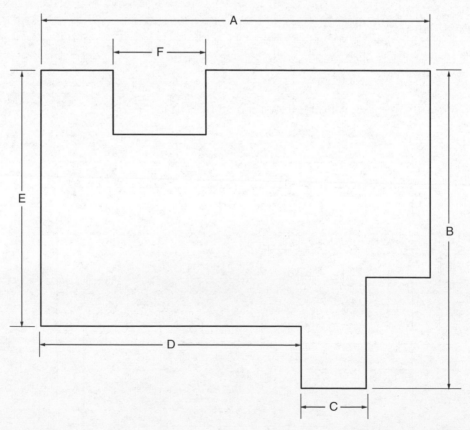

Name _____   **Date** _____   **Class** _____

113. Read and record the following distances in millimeters.

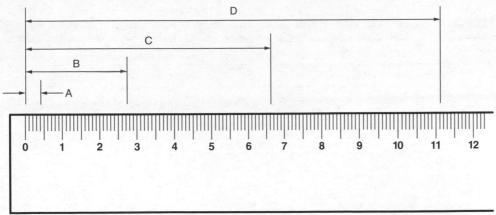

114. Using a straightedge, draw precise, sharp lines of the following lengths.

A. 102 mm   D. 133 mm

B. 12 mm   E. 44 mm

C. 165 mm   F. 90 mm

Refer to the following diagram and perform calculations to fill in the blanks in the table.

Prepare a list of decimal equivalents of all the dimensions from the preceding diagram. Round each decimal equivalent to three decimal places.

Prepare a list of the metric equivalents in millimeters of all the dimensions from the preceding diagram. Round each metric decimal equivalent to two decimal places.

	Fractional Dimensions	Decimal Equivalents	Metric Equivalents
115.	7/16"		
116.	29/64"		
117.	1/2"		
118.	33/64"		
119.	17/32"		
120.	9/16"		
121.	4 31/32"		
122.	17 3/4"		

Goodheart-Willcox Publisher

Review the following diagram. Using a tolerance of –0.5 mm, calculate the maximum and minimum allowable dimensions. Record your answers in the table provided.

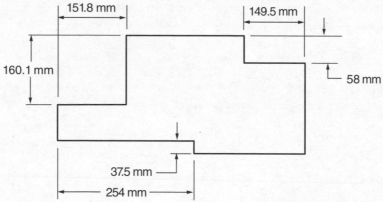

Goodheart-Willcox Publisher

	Decimal Dimensions	Maximum	Minimum
123.	37.5 mm		
124.	58.0 mm		
125.	149.5 mm		
126.	151.8 mm		
127.	160.1 mm		
128.	254.0 mm		

Goodheart-Willcox Publisher

Name _____ **Date** _____ **Class** _____

Review the following diagram. Specified lengths are designated with a letter. Using a metric scale, measure each of these lengths to the nearest millimeter. Record these values in the Metric Decimal Measurements column of the table provided.

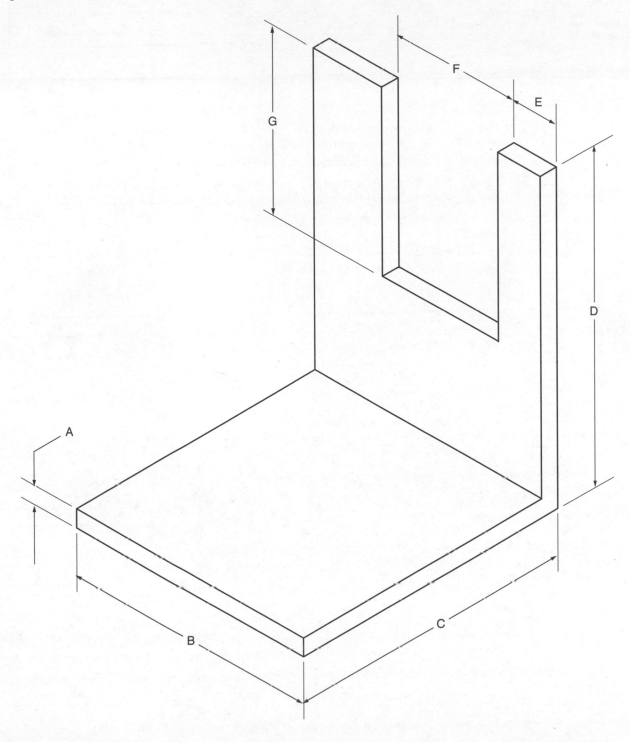

	Letter Designations	Metric Decimal Measurements	Equivalent Decimal Inches	Equivalent Fractional Inches
129.	A			
130.	B			
131.	C			
132.	D			
133.	E			
134.	F			
135.	G			

Goodheart-Willcox Publisher

Calculate the equivalent dimensions in decimal inches to the nearest hundredth of an inch. Show your work. Record these calculated values in the Equivalent Decimal Calculations column of the table.

Calculate the equivalent dimensions in fractional inches to the nearest sixty-fourth of an inch. Record these calculated values in the Equivalent Fractional Calculations column of the table.

UNIT 16

Angular Measure

Key Terms

angle

degree

minute

protractor

second

Introduction

An *angle* can be defined as the opening between two intersecting lines.

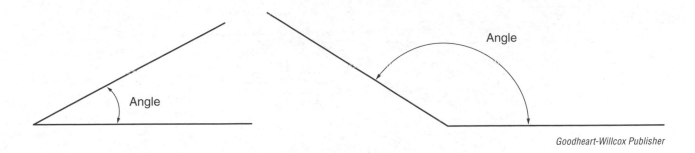

Welders are often called on to work with angles. Therefore, you should be able to take measurements of angles and to perform math operations involving angles.

Angular Measurement

The units of measure used for angles are degrees, minutes, and seconds. The largest unit is the degree. Mathematicians have divided the circle into 360 parts, calling each part a *degree*. Therefore, 1 degree is 1/360 of a circle. The notation for degrees is °. To provide a more accurate measurement, each degree has been divided into 60 equal parts, called *minutes*, and noted as '. To allow for even more accurate measurement, each minute has been divided into 60 equal parts called *seconds*, and noted as ". (Unfortunately, the notation for minutes and seconds is the same as that used for feet and inches in linear measure.)

Angles can be measured using an instrument called a *protractor*.

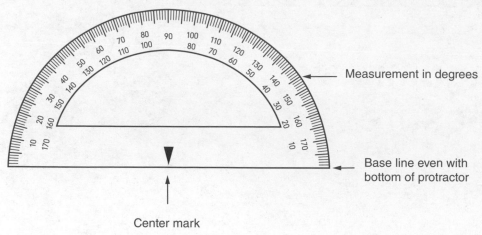

Measurement in degrees

Base line even with bottom of protractor

Center mark

To use a protractor, align the base of the protractor along one line of the angle. Position the protractor's center marker at the intersection of the lines making the angle. Then, follow the line of the angle that does not run along the base of the protractor. This line will cross under the curved part of the protractor, showing the number value of the angle. In the following example, the protractor shows the angle as 55°.

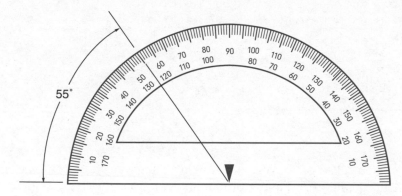

55°

Since the degree of accuracy in measuring angles is limited by the tools being used and the material being measured, your shop work is normally allowed a tolerance. As with linear measure, the tolerance may be +, –, or ±. A typical example on a blueprint may call for an angle of 53° ± 2°. In this case, the angle could vary from 55° to 51° and still be acceptable.

Angular Calculation

The basic math operations of addition, subtraction, multiplication, and division can be performed with angles.

Addition of Angles

When adding angles, align the units of measure (degrees, minutes, seconds) in columns. Add each unit, starting from the far right.

$$
\begin{array}{r r r}
45° & 39' & 17'' \\
+\,18° & 55' & 40'' \\
\hline
63° & 94' & 57''
\end{array}
$$

In many cases, the answer will produce minutes or seconds of 60 or more. In the previous example, the minutes total 94. Since 60′ are equal to one degree, add one degree to the total degrees and remove 60′ from the total minutes.

$$
\begin{array}{r r r}
63° & 94' & 57'' \\
-\,0° & 60' & 0'' \\
\hline
63° & 34' & 57'' \\
+\,1° & 0' & 0'' \\
\hline
64° & 34' & 57''
\end{array}
$$

The final answer is 64° 34′ 57″. If the seconds were 60 or greater, one minute would be added to the minutes and 60 seconds removed from the seconds.

Subtraction of Angles

As in addition of angles, align the units in columns for subtraction.

$$
\begin{array}{r r r}
11° & 52' & 10'' \\
-\,7° & 29' & 37''
\end{array}
$$

In this example, 37″ cannot be subtracted from 10″. Borrow 1′ from the 52′, add 60″ to the 10″.

$$
\begin{array}{r r r}
11° & 52' & 10'' \\
-\,0° & 1' & 0'' \\
\hline
11° & 51' & 10'' \\
+\,0° & 0' & 60'' \\
\hline
11° & 51' & 70''
\end{array}
$$

Now subtraction is possible without hindrance.

$$
\begin{array}{r r r}
11° & 51' & 70'' \\
-\,7° & 29' & 37'' \\
\hline
4° & 22' & 33''
\end{array}
$$

When subtracting angles, be mindful of how many units will be borrowed between columns.

Multiplication of Angles

When multiplying angles, treat each column of units separately. Multiply each column of units by the multiplier and keep all the digits of each unit together.

$$
\begin{array}{r r r}
30° & 47' & 22'' \\
\times & & 4 \\
\hline
120° & 188' & 88''
\end{array}
$$

After multiplication is finished, reduce each column of units that has exceeded its high value limit. Begin this process from the right side of the angle number. In this example, subtract 60″ from the 88″ and add 1′ to the 188′ to arrive at 189′.

$$
\begin{array}{ccc}
120° & 188′ & 88″ \\
-\,0° & 0′ & 60″ \\
\hline
120° & 188′ & 28″ \\
+\,0° & 1′ & 0″ \\
\hline
120° & 189′ & 28″ \\
\end{array}
$$

Notice the high value of the minute column. It takes only 60′ to equal 1°. Dividing 189 by 60 results in 3, with a remainder of 9. Since 3 is the number of times 60 fits into 189, the 3 is added to the degree column. The remainder of 9 is left over as the new minute value.

$$
\begin{array}{ccc}
120° & 189′ & 28″ \\
-\,00° & 180′ & 00″ \\
\hline
120° & 9′ & 28″ \\
+\,03° & 0′ & 0″ \\
\hline
123° & 9′ & 28″ \\
\end{array}
$$

Division of Angles

While multiplication of angles begins with the smallest unit (seconds), division of angles begins with the largest unit (degrees). Each set of units is divided separately. After division of a column of units is complete, any remainders are converted to the next lower valued column's unit and added to the existing values. Then division of that column may begin.

$$
\begin{array}{r}
\phantom{3\,\rlap{/}}\quad 40° \qquad 34′ \qquad 44″ \\
3\,\overline{)\,121° \qquad 44′ \qquad 14″} \\
\underline{-\,12\downarrow} \\
01 \longrightarrow +\,60 \\
\overline{104} \\
\underline{-\,9\downarrow} \\
14 \\
\underline{-\,12} \\
2 \longrightarrow +\,120 \\
\overline{134} \\
\underline{-\,12\downarrow} \\
14 \\
\underline{-\,12} \\
2\ r
\end{array}
$$

When a remainder occurs, round it to the nearest second. To do so, use the remainder and divisor to form the fraction 2/3. Dividing 2 by 3 approximately equals 0.67. Since 0.67 is closer to 1 than 0, the number of seconds is rounded up. The answer is 40° 34′ 45″ rounded to the nearest second.

Name _____ **Date** _____ **Class** _____

Using a protractor, measure the following angles as accurately as possible. List your answers on the lines provided.

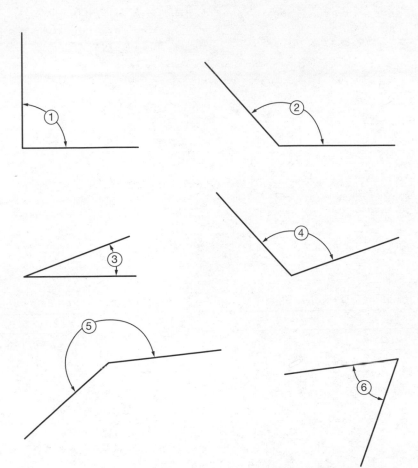

Goodheart-Willcox Publisher

1. _____   2. _____

3. _____   4. _____

5. _____   6. _____

Add the following angles. Show all your work. Be certain the columns line up. Box your answers. Using a protractor, draw each calculated angle as accurately as possible.

7. $7° \ 20'$
    $+ \ 3° \ 18'$

8. $92° \ 36' \ 10''$
    $+ \ \ \ \ \ \ 13' \ 35''$

9. $45° \ 45' \ \ 9''$
    $+ 29° \ 35' \ 51''$

10. $152° \ 32' \ 54''$
    $+ \ 27° \ 27' \ \ 6''$

11. $109°\ 51'\ 44''$
    $+\ 90°\ 47'\ 39''$

12. $91°\ 55'\ 12''$
    $+\ 76°\ \ 2'\ 14''$

13. $\ 55°\ 13'\ 16''$
    $+\ 11°\ 37'\ 39''$

14. $19°\ 50'\ \ 1''$
    $\ 4°\ \ 6'\ 19''$
    $+\ 32°\ \ 9'\ 20''$

Subtract the following angles. Show all your work. Be certain the columns line up. Box your answers. Using a protractor, draw each calculated angle as accurately as possible.

15. $119°\ 17'\ 43''$
    $-\ 91°\ 15'\ 23''$

16. $10°\ 40'$
    $-\ 7°\ 37'$

17. $135°\ 54'\ 11''$
    $-101°\ 55'\ \ 9''$

18. $204°\ \ 6'\ 32''$
    $-\ 69°\ \ 3'\ 42''$

19. $652°\ 44'\ \ 0''$
    $-359°\ \ 0'\ 18''$

20. $80°\ 22'\ 31''$
    $-\ 79°\ 31'\ 59''$

21. $45°\ \ 0'\ 21''$
    $-\ 33°\ \ 7'\ 53''$

22. $86°\ 36'\ 12''$
    $-\ \ \ \ \ \ \ 23'\ 14''$

Multiply the following angles. Show all your work. Be certain the columns line up. Box your answers.

23. $35°\ 24'\ 22''$
    $\times\ \ \ \ \ \ \ \ \ \ \ \ 8$

24. $46°\ 11'\ 48''$
    $\times\ \ \ \ \ \ \ \ \ \ \ \ 9$

25. $129°\ 58'\ 36''$
    $\times\ \ \ \ \ \ \ \ \ \ \ \ 7$

26. $90°\ 43'$
    $\times\ \ \ \ \ \ \ \ \ 5$

27. $206°\ 32'\ 47''$
    $\times\ \ \ \ \ \ \ \ \ \ \ \ 6$

28. $18°\ 14'\ \ 3''$
    $\times\ \ \ \ \ \ \ \ \ \ \ 12$

Divide the following angles. Show all your work. Be certain the columns line up. Box your answers. Using a protractor, draw each calculated angle as accurately as possible.

29. $3\overline{)393°\ \ 42'\ \ 27''}$

30. $2\overline{)181°\ \ 13'\ \ 50''}$

31. $6\overline{)180°}$

32. $8\overline{)222°}$

33. $7\overline{)360°}$

34. $8\overline{)360°}$

35. $11\overline{)93°\ \ 7'\ \ 45''}$

36. $27\overline{)19°\ \ 14'\ \ 57''}$

Name _____   **Date** _____   **Class** _____

Review the information in the following table. Calculate the maximum and minimum angles according to the given tolerance and record these values in the table.

	Angle	Tolerance	Maximum	Minimum
37.	90°	− 5°		
38.	101°	+ 30′		
39.	37°	± 3° 35′		
40.	83° 30′	± 50′		
41.	30°	± 30″		
42.	45°	+ 1° 15′		
43.	212° 47′	− 30′		
44.	66°	± 20′		

Goodheart-Willcox Publisher

Using a protractor, draw angles of the following sizes:

45. 90°   46. 28°

47. 261°   48. 180°

49. 125°   50. 350°

51. 15°   52. 110°

53. Twelve studs are to be welded together end-to-end to form a circle. What is the angle between each stud? Sketch the design including the values of each angle.

54. Sixteen evenly spaced holes are to be drilled in a round flange. What is the angle between each hole? Sketch your answer including the values of each angle.

55. What size is each angle when 180° is divided into nine equal parts?

56. A circular plate is to be divided into 15 equal parts. What is the angle of each part?

Creative Family/Shutterstock.com

Four-Sided Measure

Key Terms

area

parallelogram

perimeter

rectangle

square

square units

trapezoid

Introduction

The products that you will work on in the shop or at the job site are made up of regular geometric shapes that can be measured. Products such as frames, braces, drums, bins, hoppers, chutes, and tanks are examples of such items. In this unit, you will learn to calculate the perimeter and area of the most common four-sided shapes found in industry. The *perimeter* is the distance around a shape. The *area* is the surface measure of a shape.

Square

A *square* is a shape having four sides, with its opposite sides being parallel, all sides being the same length, and all its angles being 90°.

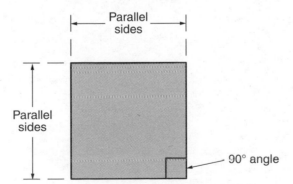

Goodheart-Willcox Publisher

Perimeter of a Square

To calculate the perimeter of a square, add the four sides.

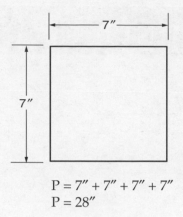

Goodheart-Willcox Publisher

$$P = 7'' + 7'' + 7'' + 7''$$
$$P = 28''$$

Area of a Square

To calculate the area of a square, multiply the two sides.

$$A = 7'' \times 7''$$
$$A = 49 \text{ in}^2$$

Area is always expressed in **square units**, such as square feet, square millimeters, square miles, etc. Remember the rules for calculating denominate numbers. When two denominate numbers are multiplied, their units combine. When those two units are the same, they become a squared unit, as seen in the example above. Review the principles of denominate numbers in Units 1–5. Common notations for area include the following:

49 square feet
49 sq/ft
49 ft^2
49 sq. ft.
750 mm^2

Rectangle

A *rectangle* is a shape having four sides, with its opposite sides being parallel and equal in length, and all its angles being 90°. It is similar to a square except one side is longer (length) than the adjacent side (width).

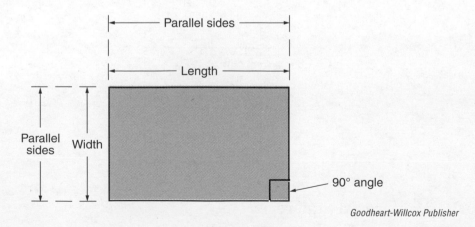

Goodheart-Willcox Publisher

Perimeter of a Rectangle

To calculate the perimeter of a rectangle, add the four sides.

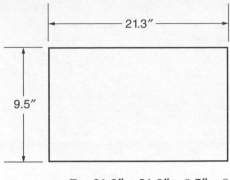

Goodheart-Willcox Publisher

$$P = 21.3'' + 21.3'' + 9.5'' + 9.5''$$
$$P = 61.6''$$

Area of a Rectangle

To calculate the area of a rectangle, multiply length by width.

$$A = 21.3'' \times 9.5''$$
$$A = 202.35 \text{ in}^2$$

Parallelogram

A *parallelogram* is a shape having four sides. Its opposite sides are the same length and are parallel. The angles opposite each other are equal. It looks like a rectangle that has been tilted.

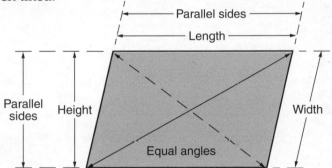

Goodheart-Willcox Publisher

Perimeter of a Parallelogram

To calculate the perimeter of a parallelogram, add the four sides.

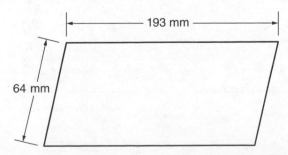

Goodheart-Willcox Publisher

$$P = 193 \text{ mm} + 193 \text{ mm} + 64 \text{ mm} + 64 \text{ mm}$$
$$P = 514 \text{ mm}$$

Area of a Parallelogram

To calculate the area of a parallelogram, multiply the length by the height.

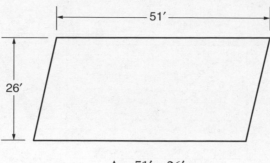

Goodheart-Willcox Publisher

$$A = 51' \times 26'$$
$$A = 1{,}326 \text{ ft}^2$$

Notice it is *not* the length and the width that are multiplied, but the length of one side and the height. You may find it interesting to note that geometric shapes are often closely related. If you were to cut the red area off the left side of the following parallelogram and then reassemble it on the right, as shown, you would create a rectangle measuring 26' × 51'. This is why the area of a parallelogram is calculated by multiplying the length and the height.

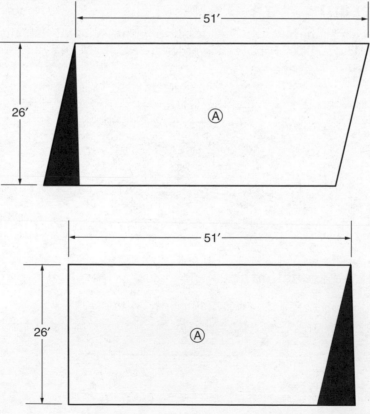

Goodheart-Willcox Publisher

Trapezoid

A *trapezoid*, like all the shapes studied so far, has four sides. Two of the sides are parallel, but the other two sides are not parallel to each other. For this reason, a trapezoid is not as symmetric or pleasing to look at as a parallelogram. Also, it does not appear in fabrication and construction as often as the parallelogram.

Goodheart-Willcox Publisher

Perimeter of a Trapezoid

To calculate the perimeter of a trapezoid, add the four sides.

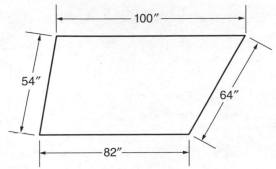

Goodheart-Willcox Publisher

$$P = 54'' + 100'' + 64'' + 82''$$
$$P = 300''$$

Area of a Trapezoid

Follow these steps to calculate the area of a trapezoid:

1. Add the two parallel sides.

$$331 + 262 = 593$$

2. Multiply by the height.

$$593 \times 204 - 120972$$

3. Divide by 2.

$$120972 \div 2 = 60486 \text{ mm}$$

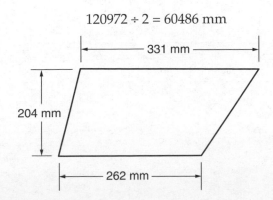

Goodheart-Willcox Publisher

As stated previously, geometric shapes are often related. If you make two trapezoids of the same size, rotate one of them 180°, and then butt them end to end, you will have formed a parallelogram that is twice the size of the original trapezoid.

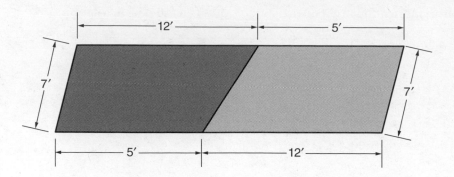

Converting Square Units of Measure

In the welding trade, the most common units of square measure are square feet, square inches, and square millimeters. Each unit of measure can be converted to the other units.

Square Feet to Square Inches

A square with sides measuring 1′ has an area of one square foot. Since 1′ equals 12″, the square also has an area of 12″ × 12″ or 144 in².

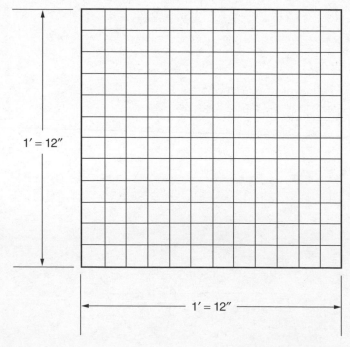

Therefore, to change square feet to square inches, multiply the number of square feet by 144.

$$1 \text{ ft}^2 = 1 \times 144 = 144 \text{ in}^2$$

Square Inches to Square Feet

To convert square inches to square feet, divide the number of square inches by 144.

1. $144 \text{ in}^2 = 144 \div 144 = 1 \text{ ft}^2$
2. $1{,}000 \text{ in}^2 = 1{,}000 \div 144 = 6.9 \text{ ft}^2$ (rounded)
3. $3{,}600 \text{ in}^2 = 3{,}600 \div 144 = 25 \text{ ft}^2$

Square Inches to Square Millimeters

A square with sides measuring 1″ has an area of one square inch. Since 1″ equals 25.4 mm, the square also has an area of 25.4 mm × 25.4 mm or 645.16 square millimeters (mm^2).

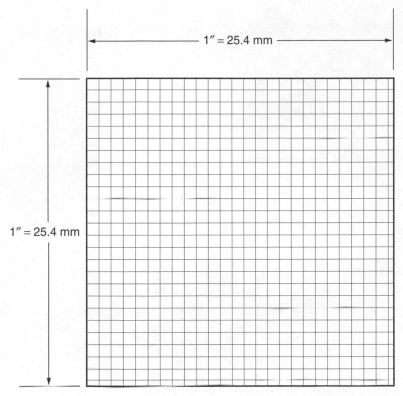

Goodheart-Willcox Publisher

Therefore, to change square inches to square millimeters, multiply the number of square inches by 645.16.

1. $1 \text{ in}^2 = 1 \times 645.16 = 645.16 \text{ mm}^2$
2. $1{,}000 \text{ in}^2 = 1{,}000 \times 645.16 = 645160 \text{ mm}^2$
3. $2{,}304 \text{ in}^2 = 2{,}304 \times 645.16 = 1486449 \text{ mm}^2$ (rounded)

Square Millimeters to Square Inches

To convert square millimeters to square inches, divide the number of square millimeters by 645.16.

1. $645.16 \text{ mm}^2 = 645.16 \div 645.16 = 1 \text{ in}^2$
2. $6000 \text{ mm}^2 = 6000 \div 645.16 = 9.3 \text{ in}^2$ (rounded)
3. $500000 \text{ mm}^2 = 500000 \div 645.16 = 775 \text{ in}^2$ (rounded)

From	To	Do
Square feet	Square inches	× 144
Square inches	Square feet	÷ 144
Square inches	Square millimeters	× 645.16
Square millimeters	Square inches	÷ 645.16

Goodheart-Willcox Publisher

Name _____ **Date** _____ **Class** _____

Convert the following square feet to square inches. Round to the nearest square inch. Show all your work. Be certain the columns line up. Box your answers.

1. 35 ft^2

2. 115 1/2 ft^2

3. 9.72 ft^2

4. 3,964 ft^2

5. 0.45 ft^2

6. 1,000 ft^2

Convert the following square inches to square feet (to the nearest tenth). Show all your work. Be certain the columns line up. Box your answers.

7. 144 in^2

8. 256 3/4 in^2

9. 15,984 in^2

10. 117.36 in^2

11. 100,000 in^2

12. 72 in^2

Convert the following square inches to square millimeters. Round to the nearest millimeter. Show all your work. Be certain the columns line up. Box your answers.

13. 90 in^2

14. 654.8 in^2

15. 137 1/8 in^2

16. 1,008 in^2

17. 1.0 in^2

18. 15.5 in^2

Convert the following square millimeters to square inches. Round to the nearest tenth. Show all your work. Be certain the columns line up. Box your answers.

19. 1550 mm^2

20. 5161.28 mm^2

21. 100 mm^2

22. 9675.5 mm^2

23. 645.16 mm^2

24. 500000 mm^2

25. A customer orders 32 pieces of the plate in the following diagram. What is the total area of the plates?

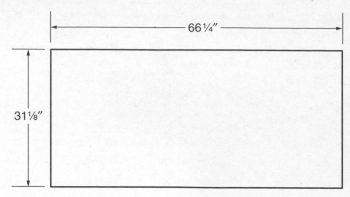

26. Calculate the perimeter of the following L-shaped plate.

27. Calculate the area of the following L-shaped plate.

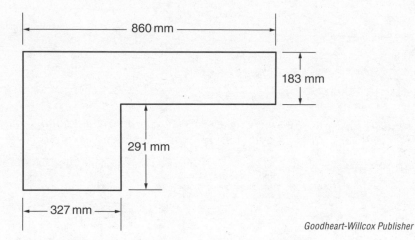

28. Calculate the area of the shaded part of the following figure in square feet.

29. Convert the area of the shaded part of the following figure from square feet to square inches.

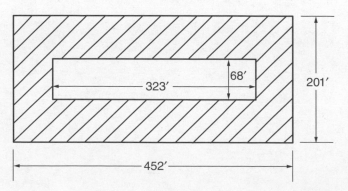

Name _____   **Date** _____   **Class** _____

30. Calculate the perimeter of the following parallelogram.

31. Calculate the area of the following parallelogram.

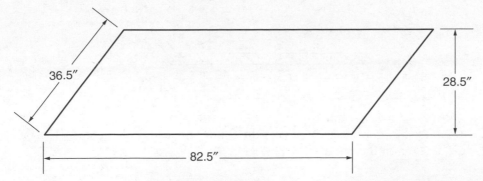

36.5″   28.5″   82.5″

Goodheart-Willcox Publisher

32. Calculate the perimeter of the following trapezoid.

33. Calculate the area of the following trapezoid.

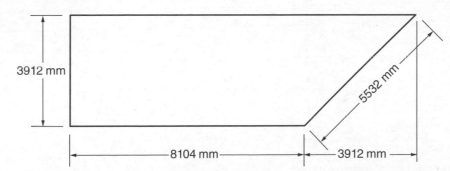

3912 mm   5532 mm   8104 mm   3912 mm

Goodheart-Willcox Publisher

34. Calculate the area of the following trapezoid.

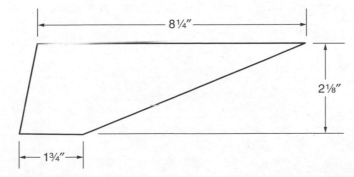

8¼″   2⅛″   1¾″

Goodheart-Willcox Publisher

35. Calculate the area of scrap remaining after the trapezoid is cut from the parallelogram.

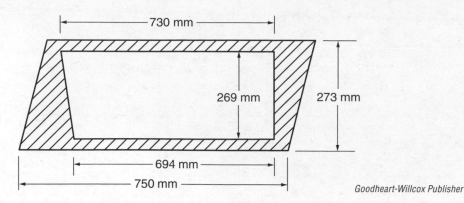

730 mm

269 mm   273 mm

694 mm

750 mm

36. A heat exchanger has 16 panels of the following size. Calculate the total surface area of the panels.

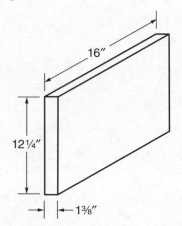

16″

12¼″

1⅜″

37. The following plate consists of a trapezoid and a parallelogram. Calculate the perimeter of the plate.

38. Calculate the area of the plate.

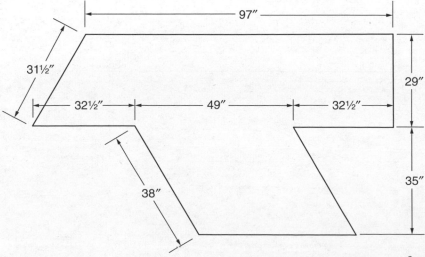

97″

31½″   29″

32½″   49″   32½″

38″   35″

Name _____ **Date** _____ **Class** _____

39. All surfaces of this sheet metal trough are to be painted with a rust-inhibiting paint. What is the total surface area, including the bottom, to be painted?

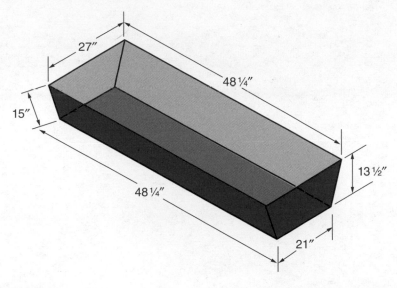

Goodheart-Willcox Publisher

Work Space/Notes

UNIT 18

Triangular Measure

Key Terms

3-4-5 triangle

equilateral triangle

hypotenuse

isosceles triangle

right triangle

triangle

Introduction

The shape explained in this unit is the three-sided figure called the *triangle*. Triangles can be divided into a number of types, but this study will be limited to the three types of triangles that appear most frequently in welding work—right, equilateral, and isosceles.

Right Triangle

A *right triangle* is a triangle in which one angle is 90°.

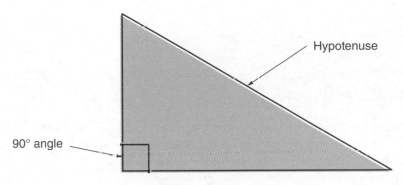

Hypotenuse

90° angle

Goodheart-Willcox Publisher

The sloping line is called the *hypotenuse*, and the other two sides are referred to as *sides*. The hypotenuse is the longest side of the triangle and is always facing the 90° angle. It is never adjacent to the 90° angle.

3-4-5 Right Triangle

Laying out a right angle is a job you may be called on to do in the shop or on the job-site. One method is to form a *3-4-5 triangle*. A triangle with hypotenuse and sides of these lengths will produce a right angle triangle. Any multiples of these numbers will work, such as 6-8-10, or 9-12-15, or 30-40-50.

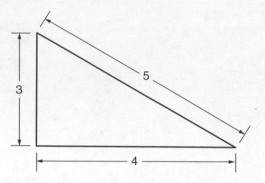

Goodheart-Willcox Publisher

Right Triangle with Equal Sides

Another method of laying out a right angle is to select two pieces of the same length for the sides, say 10'. To calculate the hypotenuse, multiply the length of one side by 1.414. In this example, the hypotenuse would be 14.14'. This hypotenuse, when assembled with the two sides, will produce a right triangle.

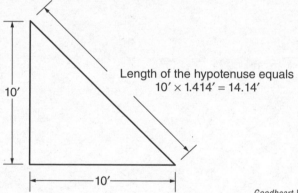

Length of the hypotenuse equals
10' × 1.414' = 14.14'

Goodheart-Willcox Publisher

Equilateral Triangle

An *equilateral triangle* is a triangle in which all three sides are the same length and all three angles are equal to 60°. An equilateral triangle is one of the strongest shapes in nature and is often used in fabricating.

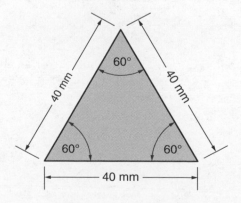

Goodheart-Willcox Publisher

Isosceles Triangle

An *isosceles triangle* is a triangle in which two of the three sides are of equal length and two of the three angles are equal.

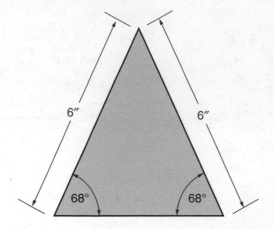

Goodheart-Willcox Publisher

Perimeter of a Triangle

To calculate the perimeter of each of these three types of triangles, simply add the three lengths or sides.

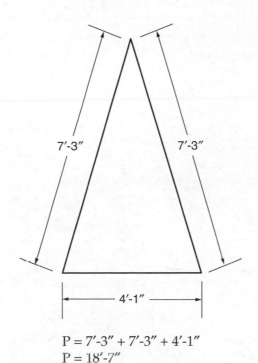

Goodheart-Willcox Publisher

P = 7'-3" + 7'-3" + 4'-1"
P = 18'-7"

Area of a Triangle

To calculate the area of each of the three types of triangles, multiply 1/2 times the base line times the height of the triangle.

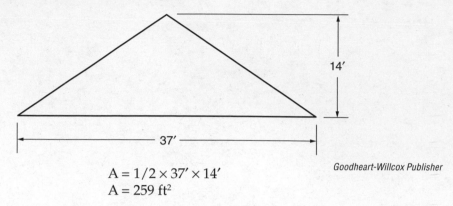

$$A = 1/2 \times 37' \times 14'$$
$$A = 259 \text{ ft}^2$$

Name _____ **Date** _____ **Class** _____

Perform the following equations as directed. Show all your work. Box your answers.

1. Calculate the perimeter of this right triangle.

2. Calculate the area of this right triangle.

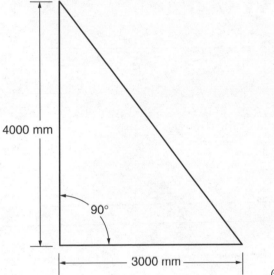

4000 mm

90°

3000 mm

Goodheart-Willcox Publisher

3. Calculate the perimeter of this isosceles triangle.

4. Calculate the area of this isosceles triangle.

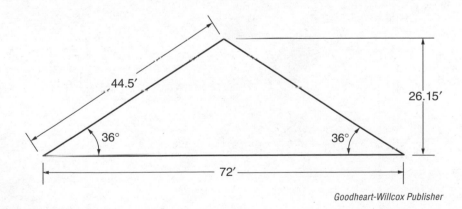

44.5′

36°

26.15′

36°

72′

Goodheart-Willcox Publisher

5. Calculate the perimeter of this equilateral triangle.

6. Calculate the area of this equilateral triangle.

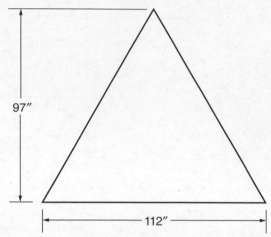

97"

112"

Goodheart-Willcox Publisher

7. A rectangular hole is cut from the following equilateral triangle. What is the remaining area of the triangle?

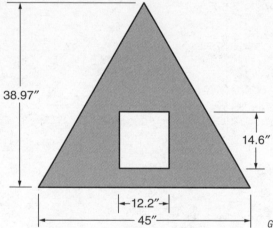

38.97"

14.6"

12.2"

45"

Goodheart-Willcox Publisher

8. What is distance Ⓐ in the following triangle?

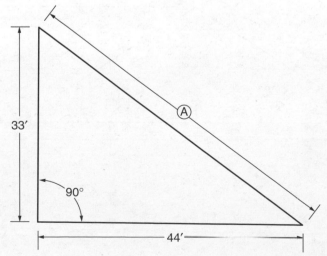

33'

Ⓐ

90°

44'

Goodheart-Willcox Publisher

Name _____   **Date** _____   **Class** _____

9. Calculate the total area of these nine equilateral triangles. Round to the nearest tenth.

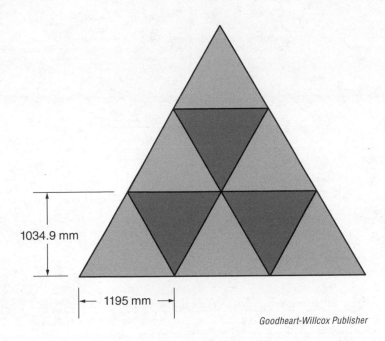

1034.9 mm

1195 mm

Goodheart-Willcox Publisher

10. Calculate the perimeter of the following triangle.

11. Calculate the area of the following triangle.

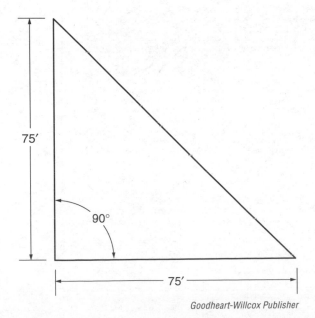

75′

90°

75′

Goodheart-Willcox Publisher

12. Review the following diagram. The centers of two bars of equal length cross at 90°. Four additional bars frame the two cross pieces. What is the total length of bar used in this piece? Express your final answer in feet and inches and round your answer to the nearest inch.

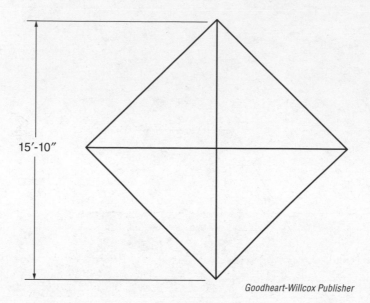

15'-10"

Goodheart-Willcox Publisher

13. Review the following diagram. What is the maximum length of tubing required for this bracket? Round your answer to the nearest 0.25".

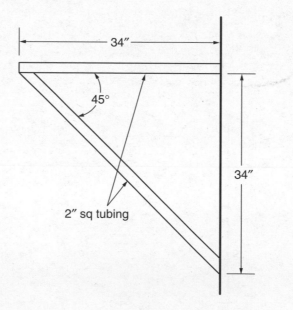

34"

45°

34"

2" sq tubing

Goodheart-Willcox Publisher

Name _____ **Date** _____ **Class** _____

14. Calculate the combined area of the triangles Ⓐ and Ⓑ in the following diagram.

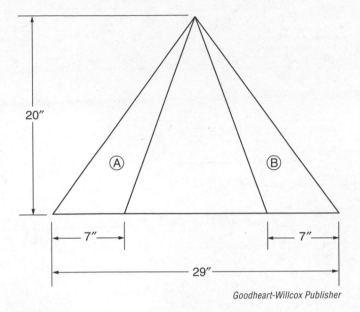

Goodheart-Willcox Publisher

Review the following diagram and use this information to answer questions 15 and 16. A plate measuring 48" × 102" was cut on an angle four times, resulting in the following shape.

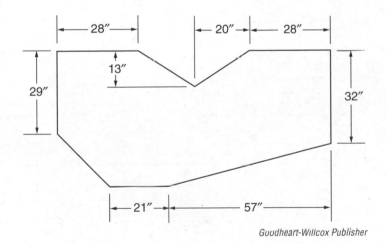

Goodheart-Willcox Publisher

15. Calculate the original area of the plate.

16. Calculate the area of the plate after cutting.

Review the following diagram and use this information to answer questions 17 and 18. The frame for the roof shown in the diagram is made of six equilateral triangles.

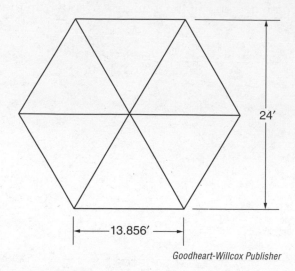

Goodheart-Willcox Publisher

17. What is the total length of beam used?

18. What is the total area of the roof?

Review the diagram of this structural frame. Answer questions 19 and 20.

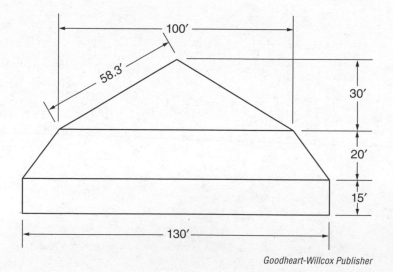

Goodheart-Willcox Publisher

Name _____   **Date** _____   **Class** _____

19. Calculate the total length of framing required.

20. Calculate the total area of the shape.

21. Review the following diagram. What is the total length of angle iron used in the roof truss?

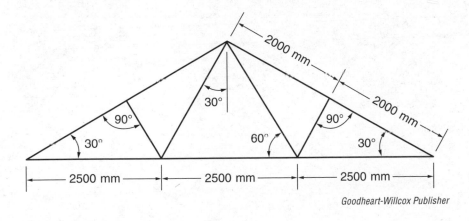

Goodheart-Willcox Publisher

Work Space/Notes

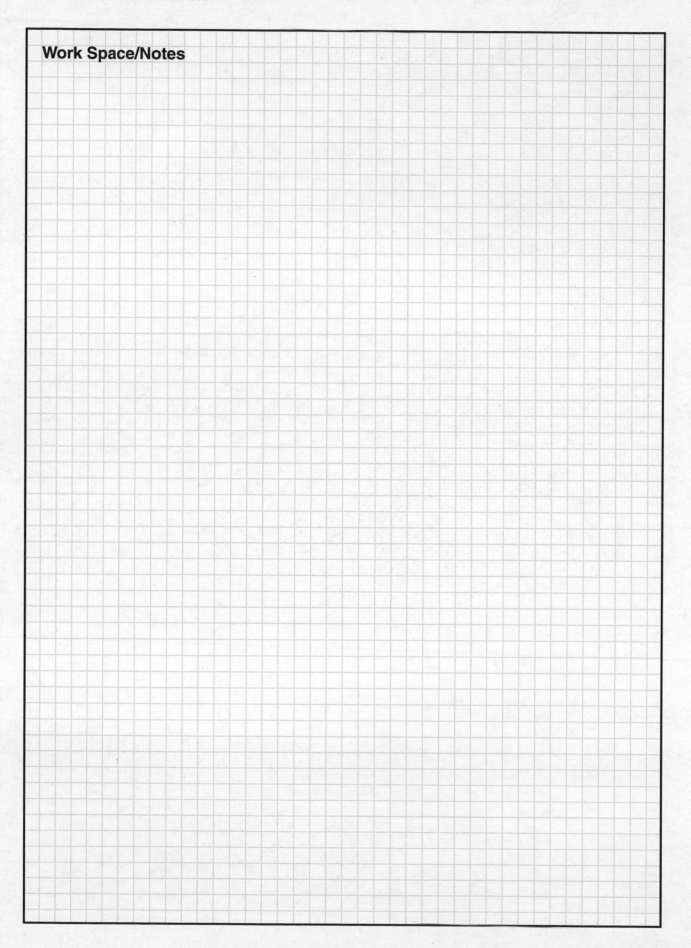

Creative Family/Shutterstock.com

UNIT 19

Circular Measure

Key Terms

circumference   radius
diameter

Introduction

Welding operations deal with objects in a variety of shapes. Working with circles and circular figures will present no problem to the welder familiar with the basic mathematical characteristics of the circle.

Circle Terms and Principles

To begin, here is the terminology you will need to know:

- The *circumference* is the distance around a circle.
- The *diameter* is the straight line distance across a circle and passing through the center.
- The *radius* is the straight line distance from the center to the edge of a circle.

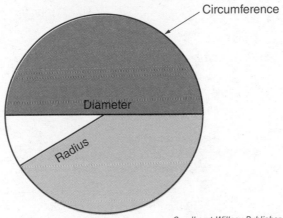

Goodheart-Willcox Publisher

If you know certain characteristics of a circle, you can calculate other characteristics.

Circumference of a Circle

Circumference can be thought of as the perimeter of a circle. If you know the diameter of a circle, the circumference can be calculated from the following formula:

$$\text{Circumference} = \pi \times \text{Diameter}$$
$$C = \pi \times D$$

This formula requires some explanation. The symbol π is a letter from the Greek language, and it represents the number 3.14. In English, it is spelled *pi* but is pronounced *pie*. By multiplying the diameter of a circle by this number, you will arrive at the circumference. So, the diameter and circumference are related to each other through the number π.

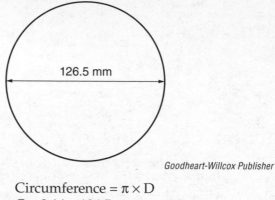

126.5 mm

Goodheart-Willcox Publisher

$$\text{Circumference} = \pi \times D$$
$$C = 3.14 \times 126.5 \text{ mm}$$
$$C = 397.21 \text{ mm}$$

Diameter of a Circle

If the circumference is known, the diameter of a circle can be calculated from the following formula:

$$D = \frac{C}{\pi}$$

$$\text{Diameter} = \frac{\text{Circumference}}{\pi}$$

Circumference = 37′

Goodheart-Willcox Publisher

$$\text{Diameter} = \frac{C}{\pi}$$

$$D = \frac{37′}{3.14}$$

$$D = 11.8′ \text{ (rounded)}$$

Area of a Circle

The area of a circle can be calculated if the radius is known.

$$\text{Area} = \pi \times \text{Radius} \times \text{Radius}$$
$$A = \pi \times R \times R$$

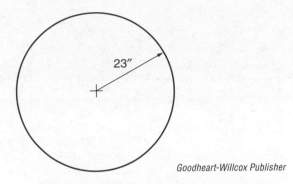

Goodheart-Willcox Publisher

$$\text{Area} = \pi \times R \times R$$
$$A = 3.14 \times 23'' \times 23''$$
$$A = 1{,}661 \text{ in}^2 \text{ (rounded)}$$

The formulas for circumference, diameter, and area are extremely important. Memorize these formulas.

Circular Shapes

Many fabricated objects consist of partial circles and straight lines. By recognizing basic shapes within these objects, you can calculate some important characteristics. The three examples of a half circle, a cylinder, and a semicircular sided shape will give you some guidance.

Half Circle

Given the information in the following diagram, the perimeter and area can be calculated.

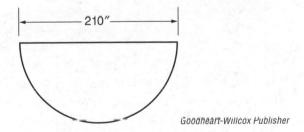

Goodheart-Willcox Publisher

Perimeter of a Half Circle

Examining a half circle reveals that its perimeter consists of the circle's diameter and half of the circle's circumference. Knowing only the diameter, its perimeter can be calculated. This is done by adding the diameter and half of the circle's circumference.

$$\text{Perimeter} = D + \frac{C}{2}$$

In the example shown, we first need to calculate the circumference of the circle. Then, divide that value in half to get the circumference of a half circle.

$$\text{Circumference (half circle)} = \pi \times \frac{D}{2}$$

$$C = \frac{3.14 \times 210''}{2} = 329.7''$$

Now we can use the circumference value in the perimeter formula.

$$\text{Perimeter (half circle)} = 329.7'' + 210'' = 539.7''$$

Area of a Half Circle

The area of a half circle is calculated by dividing in half the area of a whole circle.

$$\text{Area (half circle)} = \frac{\pi \times R \times R}{2}$$

Again, we will use the previous example with the 210″ diameter. Since a radius is simply half a circle's diameter, this radius is 105″.

$$A = \frac{3.14 \times 105'' \times 105''}{2}$$

$$A = \frac{34{,}618.5 \text{ in}^2}{2}$$

$$\text{Area (half circle)} = 17{,}309.25 \text{ in}^2$$

Cylinder

Given the information in the following diagram, the entire area of the surface of a cylinder can be calculated. This process is broken into parts. First, calculate the area of the curved surface. Next, calculate the area of the top and bottom of the cylinder. Then, add those values to the area of the curved surface of the cylinder.

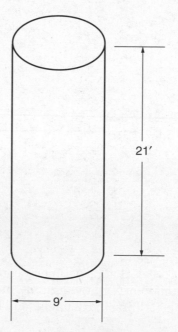

21′

9′

Goodheart-Willcox Publisher

Area of a Curved Surface

The curved surface of a cylinder is merely a rectangle wrapped as a circle. The width of the rectangle is the height of the cylinder. The length of the rectangle is the circumference of the circular end of the cylinder. Therefore, we calculate the circumference, redraw the curved surface as a flat rectangle, and use the area of a rectangle formula: Length × Width.

$$\text{Circumference} = \pi \times \text{Diameter}$$
$$C = 3.14 \times 9' = 28.26'$$

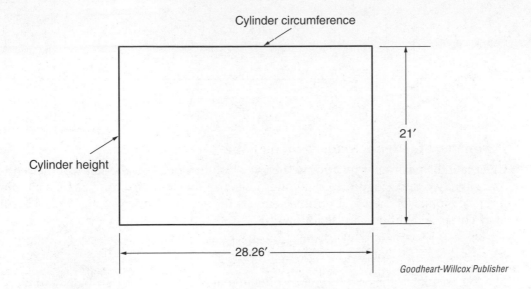

Cylinder circumference

Cylinder height

21′

28.26′

Now we have all the numbers we need to calculate the area of the curved surface of the cylinder. Plug these into the area formula.

$$\text{Area (rectangle)} = \text{Length} \times \text{Width}$$
$$\text{Area} = 21' \times 28.26' = 593.46 \text{ ft}^2$$

Area of a Cylinder

To calculate the area of the entire surface of a cylinder, add the area of the two circular ends to the area of the curved surface. Begin by calculating the area of one of the circular ends. Use the area of a circle formula.

$$\text{Area} = \pi \times \text{Radius} \times \text{Radius}$$

Since the radius is half the diameter, divide the diameter of 9′ in half to get 4.5′.

$$A = 3.14 \times 4.5' \times 4.5'$$
$$A = 63.585 \text{ ft}^2$$

Since a cylinder has two circular ends, double the area value calculated above.

$$\text{Area of cylinder ends} = 63.585 \text{ ft}^2 \times 2 = 127.17 \text{ ft}^2$$

Add the area of the curved surface of the cylinder to the area of the cylinder ends to get the area of the entire surface of the cylinder.

$$\text{Area of the entire surface of a cylinder} =$$
$$\text{Area of the curved surface} + \text{Area of the}$$
$$\text{cylinder ends}$$
$$\text{Area} = 593.46 \text{ ft}^2 + 127.17 \text{ ft}^2 = 720.63 \text{ ft}^2$$

Semicircular Sided Shape

Examine semicircular sided shapes to see of which basic shapes they are composed. This information provides clues about what formulas they require. In the following diagram, enough information is present for perimeter and area to be calculated.

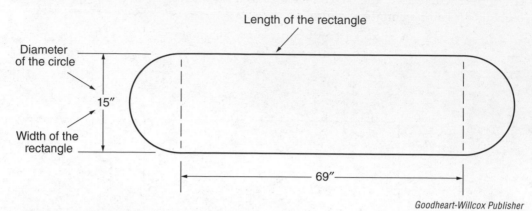

Goodheart-Willcox Publisher

Perimeter of a Semicircular Sided Shape

Calculate the perimeter by adding the two straight sides and the semicircular ends. Since the two ends together equal one circle, use the equation for a circle's circumference. Note that the height of the shape is the same distance as the circle's diameter. These calculations are as follows:

$$\text{Circumference} = \pi \times \text{Diameter}$$
$$C = 3.14 \times 15'' = 47.1''$$

Having calculated the circumference, plug in that number to the perimeter equation.

$$\text{Perimeter} = \text{Length} + \text{Length} + \text{Circumference}$$
$$\text{Perimeter} = 69'' + 69'' + 47.1'' = 185.1''$$

Area of a Semicircular Sided Shape

The area consists of two semicircles (one complete circle) plus a rectangle in the center. Begin with the area of the circle. Since this formula requires the value of the radius, calculate that number by dividing the diameter in half.

$$\text{Radius} = \frac{\text{Diameter}}{2}$$

$$\text{Radius} = \frac{15''}{2} = 7.5''$$

Use this value of the radius in the area of a circle formula.

$$\text{Area of a circle} = \pi \times R \times R$$
$$\text{Area of a circle} = 3.14 \times 7.5'' \times 7.5''$$
$$\text{Area of a circle} = 176.625 \text{ in}^2$$

Next, calculate the area of the rectangle.

$$\text{Area of rectangle} = \text{Length} \times \text{Width}$$
$$\text{Area of rectangle} = 69'' \times 15'' = 1,035 \text{ in}^2$$
$$\text{Area of semicircular sided shape} =$$
$$176.625 \text{ in}^2 + 1,035 \text{ in}^2 = 1,211.625 \text{ in}^2$$

Name _____ **Date** _____ **Class** _____

Show all your work. Box your answers.

1. Review the following diagram. Calculate the area of scrap metal remaining after the two circular parts are cut from this piece of sheet metal. Round your answer to the nearest inch.

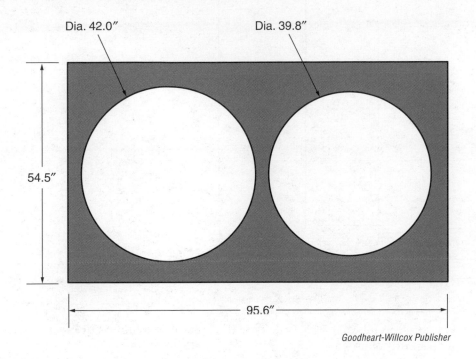

Dia. 42.0″   Dia. 39.8″

54.5″

95.6″

Goodheart-Willcox Publisher

2. Review the following diagram. Calculate the area to the nearest square foot.

3. Calculate the perimeter to the nearest inch.

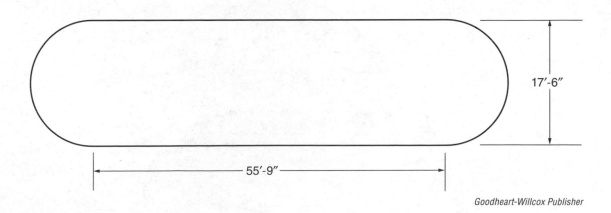

17′-6″

55′-9″

Goodheart-Willcox Publisher

4. Review the following diagram. Calculate the length of strapping needed for this pipe hanger to the nearest sixteenth of an inch.

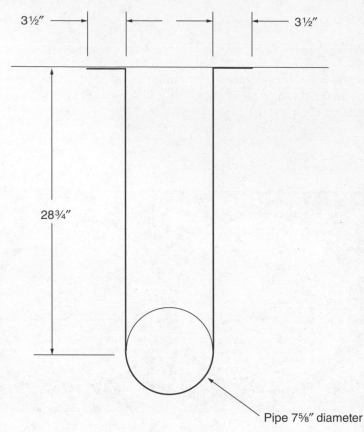

Goodheart-Willcox Publisher

5. Review the following diagram. Calculate the area of this flame-cut piece to the nearest sixteenth of an inch.

6. Calculate the perimeter of this flame-cut piece to the nearest sixteenth of an inch. A 90° segment has been removed.

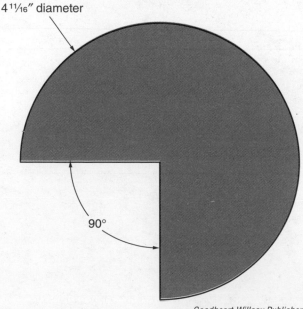

Goodheart-Willcox Publisher

Name _____   **Date** _____   **Class** _____

7. Review the following diagram. Calculate the area of the plate before the holes have been cut. Calculate to the nearest square inch.

8. Calculate the area of the plate after the holes have been cut. Calculate to the nearest square inch.

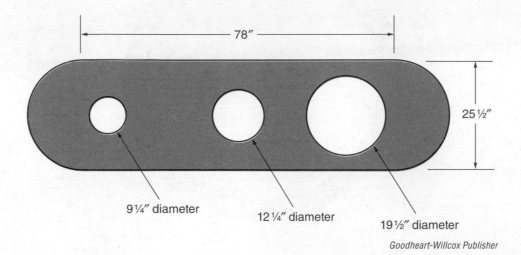

9¼" diameter

12¼" diameter

19½" diameter

Goodheart-Willcox Publisher

9. Review the following diagram. Calculate the total area of this cylinder to the nearest square inch.

10. Convert your answer from the previous question to the nearest square millimeter.

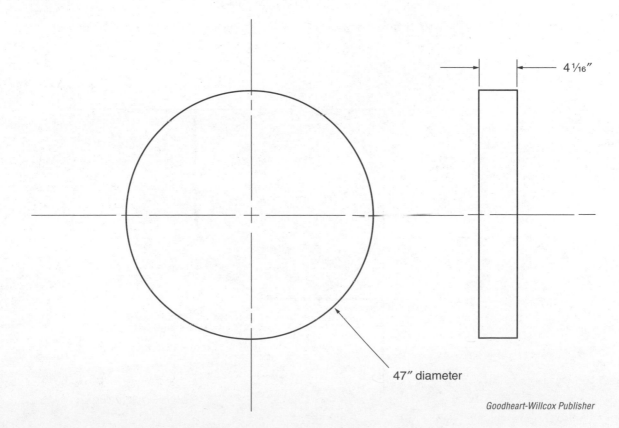

4 1/16"

47" diameter

Goodheart-Willcox Publisher

11. Calculate the area of this steel ring to the nearest square mm.

12. Convert your answer from the previous question to the nearest tenth of a square inch.

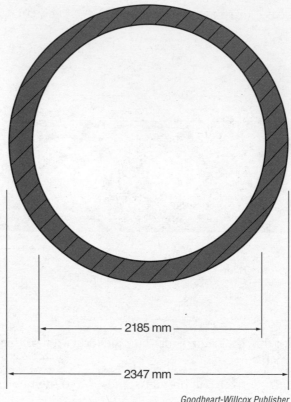

←—————— 2185 mm ——————→

←——————— 2347 mm ———————→

Goodheart-Willcox Publisher

13. Review the following diagram. Calculate the total length to the nearest tenth of an inch of strapping needed for 15 pipe corner braces.

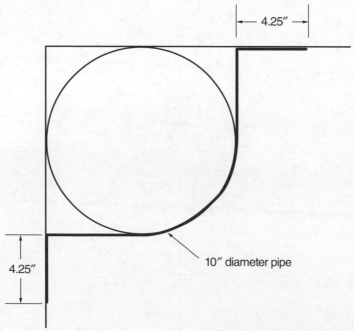

|← 4.25″ →|

4.25″

10″ diameter pipe

Goodheart-Willcox Publisher

Name _____ **Date** _____ **Class** _____

14. Review the following diagram. Calculate the total length of curved pieces needed for 39 of these wrought iron sections.

15. Calculate the total length of straight pieces needed for 39 of these wrought iron sections.

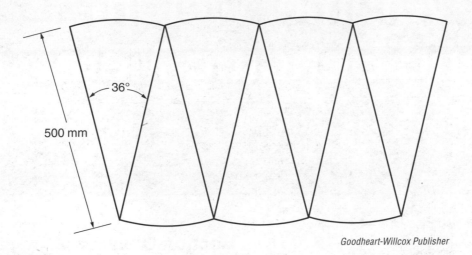

36°

500 mm

Goodheart-Willcox Publisher

16. Review the following diagram. Find the area of this shape to the nearest square inch.

17. Find the perimeter of this shape to the nearest inch.

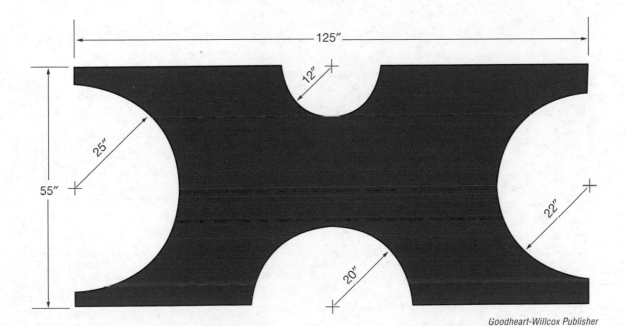

125"

12"

25"

55"

22"

20"

Goodheart-Willcox Publisher

SECTION 5
Volume, Weight, and Bending Metal

Section Objectives

After studying this section, you will be able to:

- Define regular objects.
- Calculate the volume of regular objects.
- Convert volume measurements between units.
- Calculate the volume of irregular objects.
- Compute weight calculations based on volume.
- Compute weight calculations based on length.
- Define the different types of metal bends.
- Compute flat-out length for various metal bends.

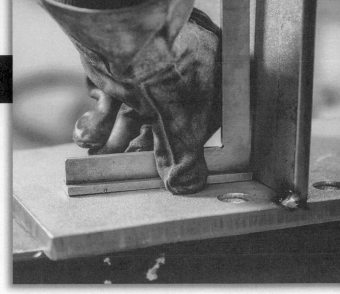

UNIT 20

Volume Measure

Key Terms

irregular object   regular object

liter   volume

Introduction

Volume is the amount of space an object occupies. All objects exist in three dimensions—length, width, and height. The volume of an object is the product of these dimensions and is expressed in cubic units. For example:

82 cubic inches
119 cubic feet
1789 mm^3
37.5 ft^3
10 in^3

Volume of Regular Objects

A *regular object* can be defined as an object having a uniform cross section. The volume of an object having a uniform cross section is determined by calculating the area of the cross section and then multiplying by the length.

$$\text{Volume} = \text{Area} \times \text{Length}$$
$$V = A \times L$$

Cube

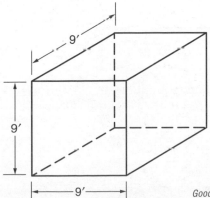

Goodheart-Willcox Publisher

Area of cross section: 9″ × 9″ = 81 in^2
Volume: 81 in^2 × 9″ = 729 ft^3

Solid Rectangle

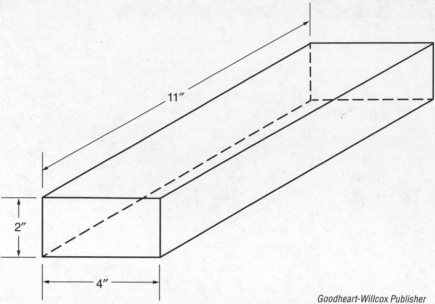

Goodheart-Willcox Publisher

Area of cross section: $2'' \times 4'' = 8$ in^2
Volume: 8 in$^2 \times 11'' = 88$ in^3

Solid Parallelogram

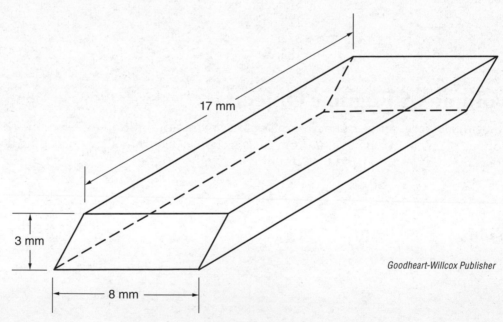

Goodheart-Willcox Publisher

Area of cross section: 3 mm $\times 8$ mm $= 24$ mm^2
Volume: 24 mm$^2 \times 17$ mm $= 408$ mm^3

Solid Trapezoid

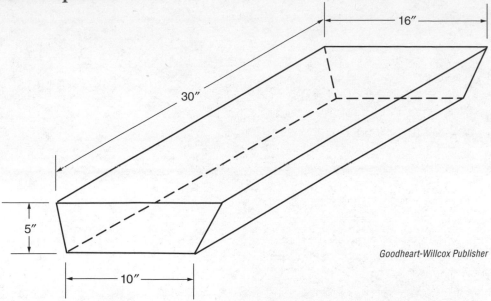

Goodheart-Willcox Publisher

Area of cross section: $16'' + 10'' = 26''$
$26'' \times 5'' = 130$ in^2
130 in$^2 \div 2 = 65$ in^2
Volume: 65 in$^2 \times 30'' = 1{,}950$ in^3

Solid Triangle

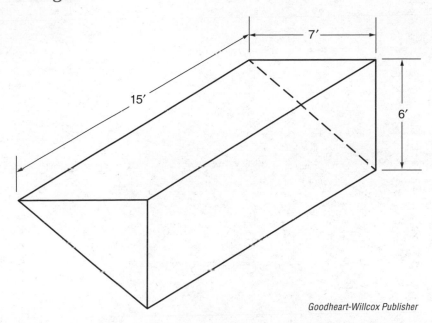

Goodheart-Willcox Publisher

Area of cross section: $7' \times 6' \times 1/2 = 21$ ft^2
Volume: 21 ft$^2 \times 15' = 315$ ft^3

Cylinder

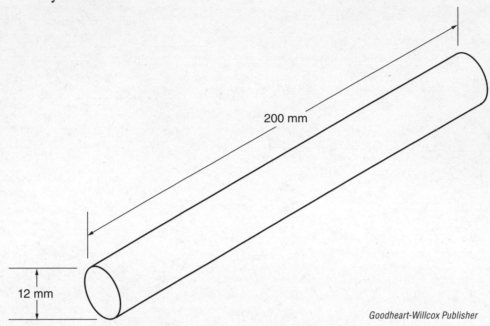

Goodheart-Willcox Publisher

Area of cross section: 3.14 × 6 mm × 6 mm = 113.04 mm²
Volume: 113.04 mm² × 200 mm = 22608 mm³

Half Cylinder

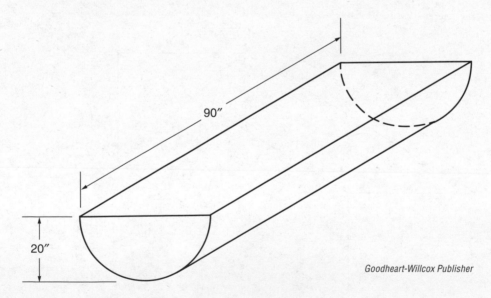

Goodheart-Willcox Publisher

Area of cross section: $\dfrac{3.14 \times 20'' \times 20''}{2} = 628 \text{ in}^2$

Volume: 628 in² × 90″ = 56,520 in³

Solid Semicircular Sided Shape

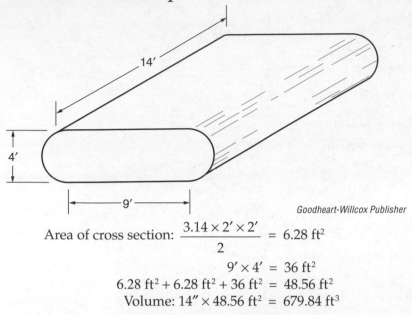

Goodheart-Willcox Publisher

Area of cross section: $\dfrac{3.14 \times 2' \times 2'}{2}$ = 6.28 ft²

$$9' \times 4' = 36 \text{ ft}^2$$
$$6.28 \text{ ft}^2 + 6.28 \text{ ft}^2 + 36 \text{ ft}^2 = 48.56 \text{ ft}^2$$
$$\text{Volume: } 14'' \times 48.56 \text{ ft}^2 = 679.84 \text{ ft}^3$$

Converting Volume Measurements

Linear dimensions of objects are usually expressed in feet, inches, or millimeters. Volume can be expressed in cubic feet, cubic inches, or cubic millimeters. However, it is often more useful to express the volume (or capacity) of an object in gallons or liters.

A *liter* is a metric measurement of capacity. A liter is a little bigger than a quart. It is important for you to be able to convert back and forth between the various units of volume. The following are examples of common equivalent volume measurements:

$$
\begin{aligned}
1 \text{ gallon (US)} &= 231 \text{ in}^3 \\
1 \text{ ft}^3 &= 7.48 \text{ gallons (US)} \\
1 \text{ gallon (US)} &= 3.785 \text{ liters} \\
1 \text{ ft}^3 &= 1{,}728 \text{ in}^3 \\
1 \text{ in}^3 &= 16387.064 \text{ mm}^3
\end{aligned}
$$

The degree of accuracy usually required in a welding shop allows the use of rounded figures.

$$
\begin{aligned}
1 \text{ in}^3 &= 16387 \text{ mm}^3 \\
1 \text{ gallon (US)} &= 3.79 \text{ liters}
\end{aligned}
$$

Also, since conversions often result in lengthy decimals, it is normal practice to round the answer.

1. Gallons to Cubic Inches:

Multiply the number of gallons by 231.
40 gallons = 40 × 231 = 9,240 in³

2. Cubic Inches to Gallons:

Divide the number of cubic inches by 231.
2,425.5 in³ = 2,425.5 ÷ 231 = 10.5 gallons

3. Gallons to Cubic Feet:

Divide the number of gallons by 7.48.
1,000 gallons = 1,000 ÷ 7.48 = 133.6898 ft³ = 133.69 ft³ (rounded)

4. **Cubic Feet to Gallons:**

Multiply the number of cubic feet by 7.48.
16 ft³ = 16 × 7.48 = 119.68 gallons

5. **Gallons to Liters:**

Multiply the number of gallons by 3.79.
5 gallons = 5 × 3.79 = 18.95 liters

6. **Liters to Gallons:**

Divide the number of liters by 3.79.
85 liters = 85 ÷ 3.79 = 22.4274 gallons = 22.43 gallons (rounded)

7. **Cubic Millimeters to Cubic Inches:**

Divide the number of cubic millimeters by 16,387.
27000000 mm³ = 27000000 ÷ 16,387 = 1,647.65 in³ (rounded)

8. **Cubic Inches to Cubic Millimeters:**

Multiply the number of cubic inches by 16,387.
9 in³ = 9 × 16,387 = 147483 mm³

9. **Cubic Feet to Cubic Inches:**

Multiply the number of cubic feet by 1,728.
27 ft³ = 27 × 1,728 = 46,656 in³

10. **Cubic Inches to Cubic Feet:**

Divide the number of cubic inches by 1,728.
5,000 in³ = 5,000 ÷ 1,728 = 2.8935 ft³ = 2.89 ft³ (rounded)

Volume of Irregular Objects

An *irregular object* can be defined as an object having a cross section that is not uniform. The volume of many irregular objects can be easily calculated. The way to compute this is to first divide the object into regular parts. Then calculate the volume of each part. Finally, add all the calculated volumes. The following example calculates the volume to the nearest tenth of an inch. The material is 2″ thick and all holes are 2″ in diameter.

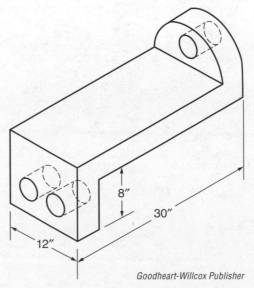

Goodheart-Willcox Publisher

1. Step One: Divide the object into regular parts.

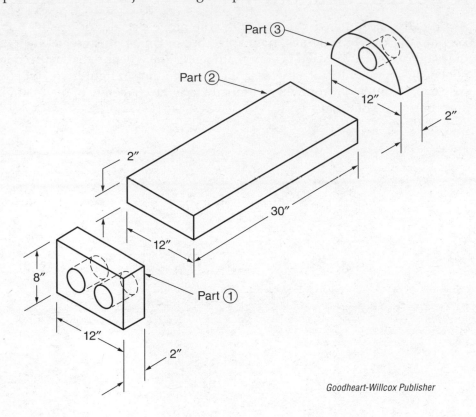

Part ③

Part ②

12″

2″

2″

30″

12″

8″

12″

2″

Part ①

Goodheart-Willcox Publisher

2. Step Two: Calculate the volume of each part.

Part 1:

Gross volume = 2″ × 8″ × 12″   =   192.00 in³
Minus volume of holes = π × 1″ × 1″ × 2″ × 2 = − 12.56 in³
Volume   =   179.44 in³

Part 2:

Volume = 2″ × 12″ × 30″ =   720.00 in³

Part 3:

$$\text{Volume of half circle} = \frac{\pi \times 6'' \times 6'' \times 2''}{2} \quad = \quad 113.04 \text{ in}^3$$

Minus volume of hole = π × 1″ × 1″ × 2″   = − 6.28 in³
Volume   =   106.76 in³

3. Step Three: Add the individual volumes together.

Part 1:   179.44 in³
Part 2:   720.00 in³
Part 3:   + 106.76 in³
Total Volume: 1,006.2 in³

A Word of Caution

A difficulty encountered in calculating volumes, especially of irregular objects, is keeping track of your figures. As pointed out in Unit 1, it is crucial that you form the habit of being organized in your math work. You will find that math concepts do not become more difficult. However, the sheer volume of figures you will deal with may cause confusion. You will continually face the challenge of maintaining order and control of your math work.

Name _____ **Date** _____ **Class** _____

Calculate the following equations. Show all your work. Be certain the columns line up.
Box your answers.

1. How many cubic inches of space does 6 1/4 gallons of water occupy?

2. How many cubic feet are contained in 29000000 mm^3? Calculate to the nearest tenth.

3. A swimming pool contains 1,600 ft^3 of water. How many gallons does the pool contain? Calculate to the nearest gallon.

4. Convert 1.125 in^3 to mm^3. Calculate to the nearest mm.

5. A home heating oil storage tank holds 480 liters of oil. How many gallons does it hold? Calculate to the nearest gallon.

6. A microwave oven measures 16 1/2″ × 11″ × 7 1/2″. What is the volume in cubic inches? Calculate to the nearest hundredth.

7. What is the volume in cubic feet of the microwave oven from the previous question? Calculate to the nearest hundredth.

8. How many liters are there in a 25 gallon gasoline tank? Calculate to the nearest tenth.

9. How many cubic inches of space are there in a refrigerator with a capacity of 8 ft³?

10. The water tank for the town of South Windsor has a capacity of 500,000 gallons. How many cubic feet of water does it contain? Calculate to the nearest cubic foot.

Name _____   **Date** _____   **Class** _____

11. A motorcycle engine has a size of 45 in³. What is the size of the engine in mm³?

12. Convert 9,000 in³ to cubic feet. Calculate to the nearest tenth.

13. A quenching tank measures 2′ × 3′ × 6′. How many gallons does it hold? Calculate to the nearest hundredth.

14. How many gallons are there in 1,000 in³? Calculate to the nearest thousandth.

15. A block of precision steel contains 2539985 mm³. What is the volume in cubic inches?

16. Convert 32,000 gallons to cubic feet. Calculate to the nearest tenth.

17. How many cubic millimeters are contained in the following piece of steel? Calculate to the nearest hundredth mm.

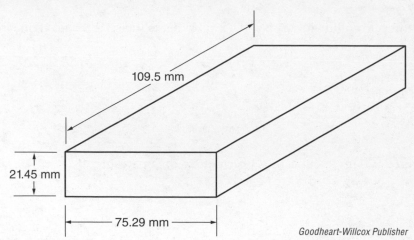

18. What is the volume of steel in this triangular rod? Calculate the answer to the nearest cubic inch.

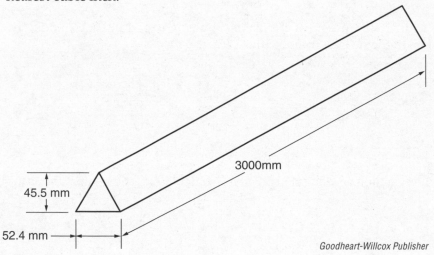

19. Review the following diagram. The concrete foundation for a newly installed press has the following dimensions. How many cubic yards of concrete were used? Calculate the answer to the nearest one quarter of a yard.

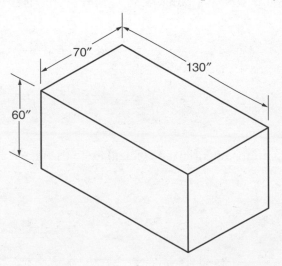

Name _____   **Date** _____   **Class** _____

20. Find the total volume of this weldment to the nearest half cubic inch.

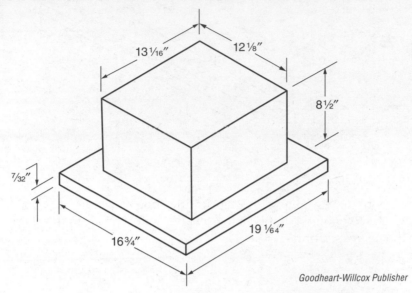

13¹⁄₁₆″   12⅛″

8½″

⁷⁄₃₂″

16¾″   19¹⁄₆₄″

Goodheart-Willcox Publisher

21. This block of steel is turned on a lathe to a diameter of 18″. What volume of material is removed? Calculate to the nearest hundredth of a cubic inch.

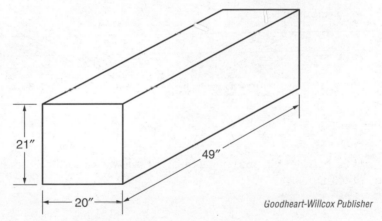

21″

49″

20″

Goodheart-Willcox Publisher

22. Review the following diagram. Calculate the volume of this shape.

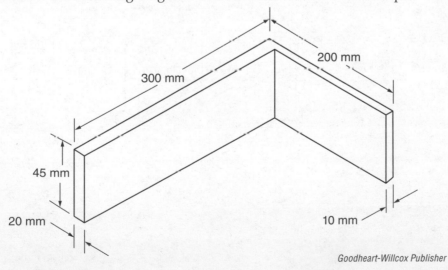

200 mm

300 mm

45 mm

20 mm   10 mm

Goodheart-Willcox Publisher

23. Fifteen joints are welded as illustrated in the following diagram. Fillet welds on both sides of the joint are similar. What volume of weld material is deposited?

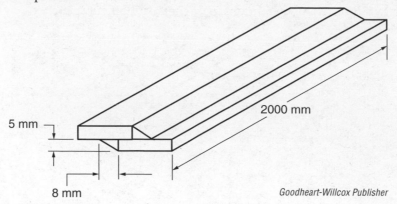

5 mm

2000 mm

8 mm

24. The ventilating system in this fabricating shop can evacuate 6,100 ft³ of air per minute. How long will it take to exchange all of the air in the shop? Calculate to the nearest minute.

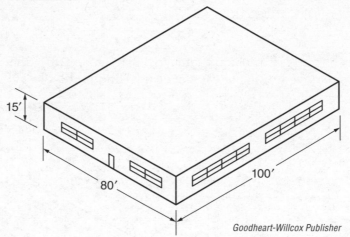

15'

80'

100'

25. What is the total volume of cast iron in this weldment? Express your answer in cubic inches.

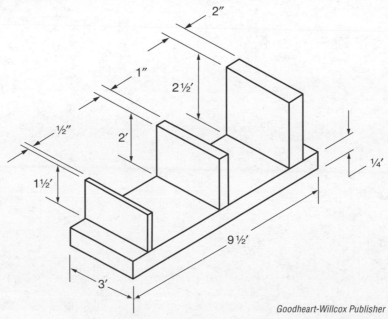

2"

1"

2½'

½"

2'

1½'

¼'

3'

9½'

Name _____   **Date** _____   **Class** _____

26. Calculate the volume of this silo to the nearest cubic foot.

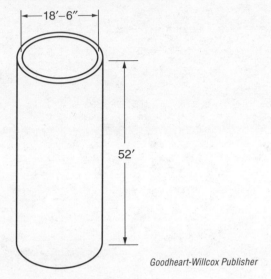

27. How many cubic feet of steel are in this 250′ roll?

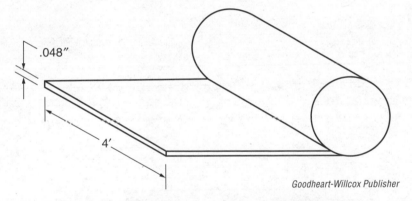

28. Nine bins are to be constructed as shown in the following diagram. What is the total volume of the bins to the nearest cubic foot?

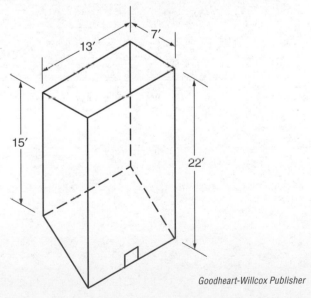

29. The worn surface of this 475 mm long shaft is built up by surfacing. Three millimeters of material are deposited all around. What total volume of weld material is deposited?

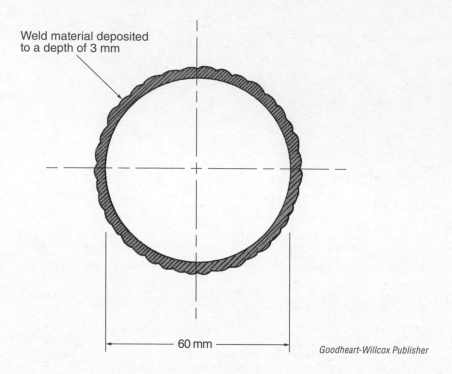

Weld material deposited
to a depth of 3 mm

60 mm

Goodheart-Willcox Publisher

30. This 17′ wide above-ground pool is filled to within 6″ of the top. How many liters of water does it contain? Calculate the answer to the nearest ten liters.

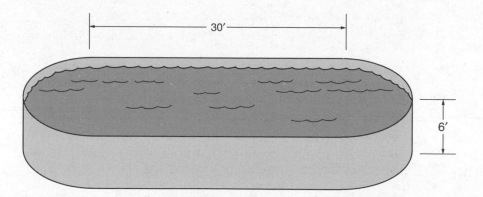

30′

6′

Goodheart-Willcox Publisher

Name _____   **Date** _____   **Class** _____

31. Review the following diagram. Which container has greater capacity—the rectangular or the half-circle container?

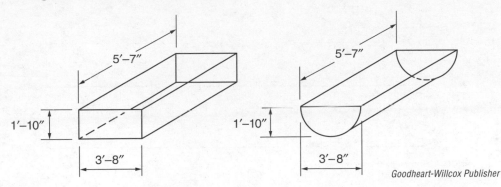

32. How many more gallons can the container with greater capacity hold? Calculate to the nearest gallon.

33. Review the following diagram. All material in this weldment is 1/4″ thick unless otherwise shown on the drawing. The flame-cut hole is 10″ in diameter. The triangular support pieces are on only one side of the upright. Find the total volume to the nearest quarter of a cubic inch.

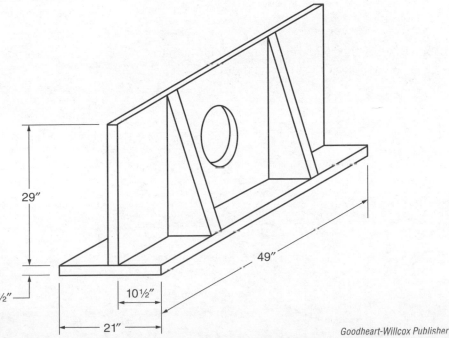

34. Review the following diagram. Calculate the volume of pipe Ⓐ to the nearest cubic inch.

35. Calculate the volume of pipe Ⓑ to the nearest cubic inch.

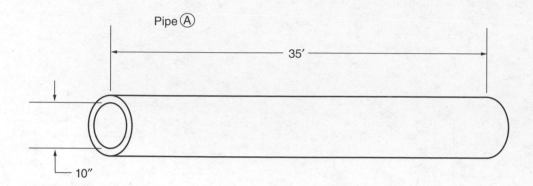

Pipe Ⓐ

35′

10″

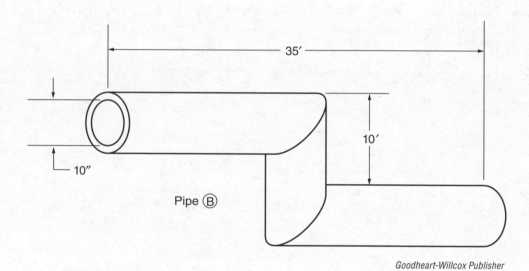

35′

10′

10″

Pipe Ⓑ

36. Review the following diagram. How many gallons will this trough hold? Calculate to the nearest gallon.

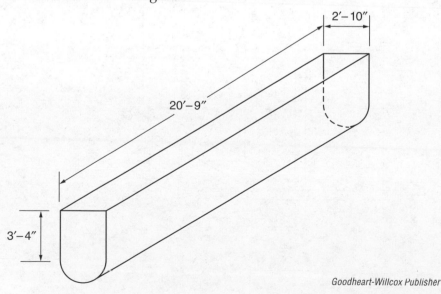

2′–10″

20′–9″

3′–4″

Name _____   **Date** _____   **Class** _____

37. Two plates, 3500 mm long, are welded together as shown. What is the total volume of weld deposited?

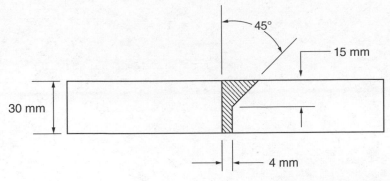

45°

15 mm

30 mm

4 mm

38. Calculate the total volume of the two settling tanks, including the pipes, to the nearest gallon.

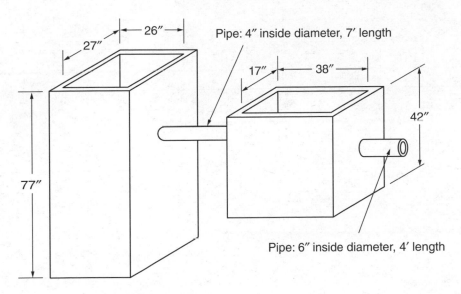

26″

27″

Pipe: 4″ inside diameter, 7′ length

17″   38″

42″

77″

Pipe: 6″ inside diameter, 4′ length

Work Space/Notes

UNIT 21

Weight Measure

Key Terms

centimeter (cm)   kilogram (kg)

gram (g)   meter (m)

Introduction

When calculating the weight of weldments, you will encounter one of the following conditions. The weldment may consist of randomly shaped metal plates, standard structural shapes (such as angle iron, channel, etc.), or a combination of both. The calculations for randomly shaped plates are based on volume. The calculations for standard structural shapes are based on length.

Weight Calculation Based on Volume

The weight of a block of steel is easy to determine if you know the volume of the piece and the weight of the steel per unit volume. For example, steel has a weight of 0.2835 lb per cubic inch. Therefore, a block of steel with a volume of 250 in³ weighs 250×0.2835, or 70.875 lb.

Metals have different densities and, therefore, different weights. Even the weight of different classifications of steel will vary depending on the composition of the steel. The following table shows the weights of various metals. The unit of measure g/cm³ means grams per cubic centimeter. Grams (g) and centimeters (cm) are metric measurements commonly used in these calculations.

Weight of Metals					
Metal	lb/in³	g/cm³	Metal	lb/in³	g/cm³
Magnesium	0.0628	1.738	Steel	0.2835	7.847
Aluminum	0.0975	2.699	Copper	0.3210	8.885
Zinc	0.2570	7.114	Lead	0.4096	11.338
Tin	0.2633	7.288	Tungsten	0.6900	19.099
Cast Iron	0.2665	7.377	Gold	0.6969	19.290

Goodheart-Willcox Publisher

The *gram (g)* is a very small measure of weight. A paper clip, for example, weighs about one gram. Since many objects handled in everyday life are much heavier than a few grams, it is common practice to use 1000 grams as a unit of measure. One thousand grams is called a ***kilogram (kg)*** and is a little more than two pounds. It can also be written as *kg* or as *kilo*. To convert from grams to kilograms, divide by 1000. To convert from kilograms to grams, multiply by 1000.

The ***centimeter (cm)*** is a measure of length and is ten millimeters long. A cube of sugar, for example, measures about 1 cm × 1 cm × 1 cm and is therefore about 1 cm³ in volume.

Because the millimeter is so small, many objects are measured in centimeters. To convert from millimeters to centimeters, divide by 10. To convert from centimeters to millimeters, multiply by 10.

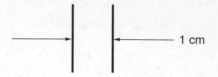

Goodheart-Willcox Publisher

Weight Calculation Based on Length

A full description of the standard structural shapes produced by American steel makers is provided in the *Steel Construction Manual* issued by the American Institute of Steel Construction. The Canadian equivalent is produced by the Canadian Institute of Steel Construction. Among the statistics provided for each shape is the weight per unit length. A summary of selected shapes is shown in the following figure. Descriptions of C shape, L shape, S shape, and W shape can be found in the Glossary.

You will notice that the figures may be expressed in the US customary system as pounds per foot or in the metric system as kilograms per meter. As previously explained, 1 kilogram equals 1000 grams. Similarly, 1 ***meter (m)*** is equal to 1000 millimeters and is a bit longer than 1 yard. You will see it written as *m*. To convert from millimeters to meters, divide by 1000. To convert from m to mm, multiply by 1000.

The weight of a standard structural shape is easy to determine if you know the length of the piece and the weight of the shape per unit length. For example, if a 4" channel weighs 5.4 lb per ft, the weight of a 6' piece would be 5.4 × 6 or 32.4 lb.

Review the following standard shapes and their weights at different lengths.

W Shapes – metric

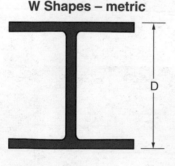

Designation D (mm) × weight (kg/m)	Weight kg/m
W 310 × 202	202
W 250 × 167	167
W 250 × 67	67
W 200 × 100	100
W 200 × 22	22
W 150 × 24	24
W 130 × 28	28

W Shapes – standard

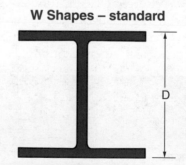

Designation D (inches) × weight (lb/ft)	Weight lb/ft
W 16 × 57	57
W 16 × 26	26
W 12 × 35	35
W 8 × 28	28
W 6 × 25	25
W 6 × 9	9
W 5 × 19	19

S Shapes – metric

Designation D (mm) × weight (kg/m)	Weight kg/m
S 610 × 158	158
S 250 × 52	52
S 200 × 34	34
S 180 × 30	30
S 150 × 26	26
S 130 × 22	22
S 75 × 11	11

S Shapes – standard

Designation D (inches) × weight (lb/ft)	Weight lb/ft
S 18 × 54.7	54.7
S 12 × 50	50.0
S 12 × 40.8	40.8
S 8 × 23	23.0
S 8 × 18.4	18.4
S 6 × 17.25	17.25
S 3 × 5.7	5.7

Channels (C Shapes) – metric

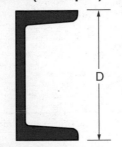

Designation D (mm) × weight (kg/m)	Weight kg/m
C 250 × 37	37
C 200 × 28	28
C 180 × 18	18
C 150 × 19	19
C 130 × 17	17
C 130 × 13	13
C 100 × 9	9

Channels (C Shapes) – standard

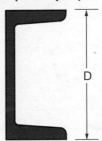

Designation D (inches) × weight (lb/ft)	Weight lb/ft
C 10 × 30	30.0
C 9 × 13.4	13.4
C 8 × 18.75	18.75
C 8 × 11.5	11.5
C 6 × 10.5	10.5
C 5 × 9	9.0
C 4 × 5.4	5.4

Angles (L Shapes) – metric

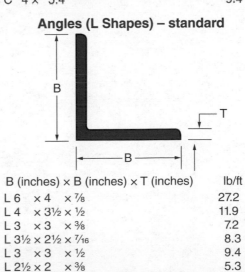

B (mm) × B (mm) × T (mm)	kg/m
L 200 × 200 × 16	48.216
L 150 × 100 × 13	24.257
L 125 × 90 × 13	20.685
L 100 × 100 × 16	23.685
L 90 × 75 × 10	12.173
L 55 × 55 × 8	6.399
L 45 × 55 × 6	3.244

Angles (L Shapes) – standard

B (inches) × B (inches) × T (inches)	lb/ft
L 6 × 4 × ⅞	27.2
L 4 × 3½ × ½	11.9
L 3 × 3 × ⅜	7.2
L 3½ × 2½ × ⁷⁄₁₆	8.3
L 3 × 3 × ½	9.4
L 2½ × 2 × ⅜	5.3
L 2 × 2 × ⅛	1.65

Goodheart-Willcox Publisher

Work Space/Notes

Name _____ **Date** _____ **Class** _____

Perform the following calculations as directed. Show all your work. Box your answers.

1. Calculate the weight of this steel bar to the nearest kilogram. (Hint: Since the table of structural steel shapes provides the weight in g/cm³, it is best to calculate the volume in cubic centimeters. First, convert the millimeter dimensions to centimeters and then proceed with finding the volume in cubic centimeters.)

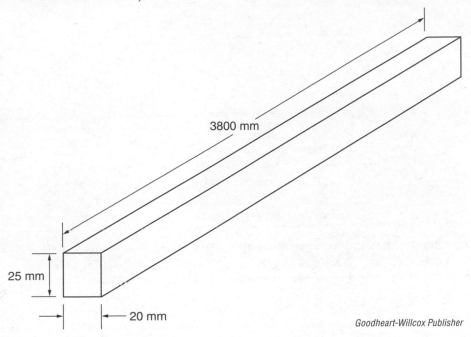

Goodheart-Willcox Publisher

2. Review the following diagram. Calculate the weight of the copper plate to the nearest pound.

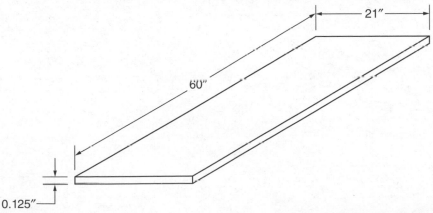

Goodheart-Willcox Publisher

3. Review the following diagram. Calculate the weight of the cast iron bar to the nearest pound.

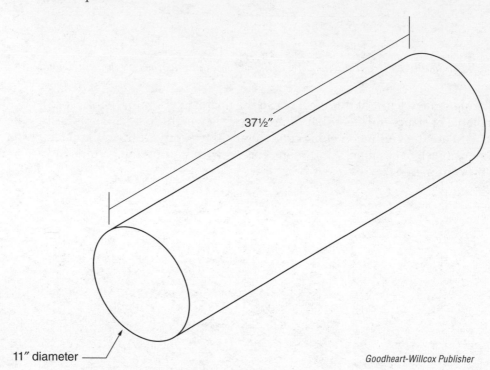

37½"

11" diameter

Goodheart-Willcox Publisher

4. Review the following diagram. Calculate the weight of the piece of angle iron to the nearest tenth of a kilogram. (NOTE: For standard structural shapes, refer to the description provided in the text.)

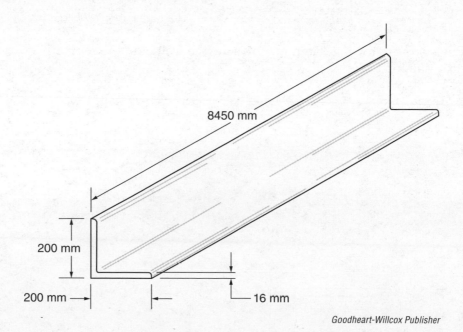

8450 mm

200 mm

200 mm

16 mm

Goodheart-Willcox Publisher

Name _____   **Date** _____   **Class** _____

5. Calculate the weight of the following S shape beam to the nearest tenth of a kilo.

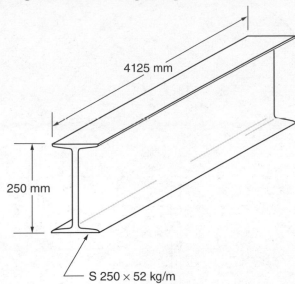

4125 mm

250 mm

S 250 × 52 kg/m

Goodheart-Willcox Publisher

6. Calculate the weight of the following triangular piece of steel to the nearest tenth of a kilogram.

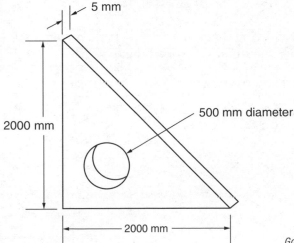

5 mm

2000 mm

500 mm diameter

2000 mm

Goodheart-Willcox Publisher

7. Calculate the weight of the following steel tube to the nearest pound.

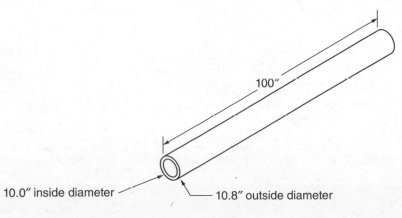

100″

10.0″ inside diameter

10.8″ outside diameter

Goodheart-Willcox Publisher

8. Calculate the weight of this piece of steel to the nearest tenth of a pound. The diameter is 40″ and the thickness is 2″.

9. Review the following diagram of a gold bar. Calculate the weight of this bar of gold to the nearest hundredth of a pound.

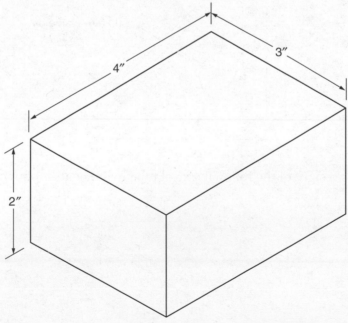

Name _____   **Date** _____   **Class** _____

10. Calculate the weight of this steel hub and plate to the nearest tenth of a pound.

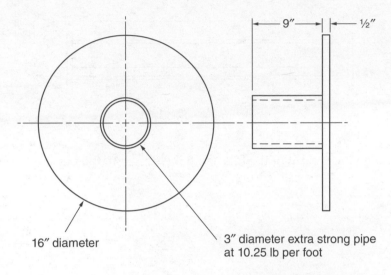

16" diameter

3" diameter extra strong pipe
at 10.25 lb per foot

Goodheart-Willcox Publisher

11. The following frame is made of square structural tubing weighing 47.9 lb per foot. What is the weight of the frame to the nearest tenth of a pound?

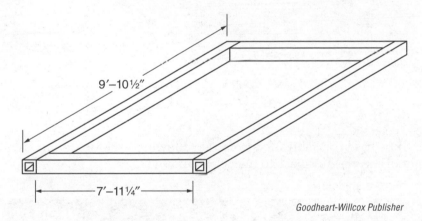

9'–10½"

7'–11¼"

Goodheart-Willcox Publisher

12. Calculate the weight of the following steel platform to the nearest kilogram.

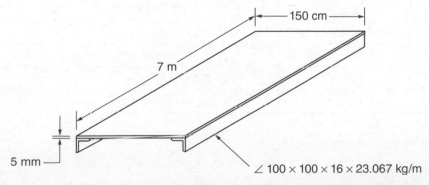

150 cm

7 m

5 mm

∠ 100 × 100 × 16 × 23.067 kg/m

Goodheart-Willcox Publisher

13. A customer order required 13 of the columns illustrated. What is the total weight of the order to the nearest pound?

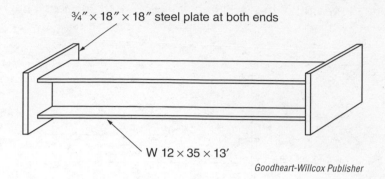

¾″ × 18″ × 18″ steel plate at both ends

W 12 × 35 × 13′

Goodheart-Willcox Publisher

14. Calculate the weight of the following weldment to the nearest kilogram.

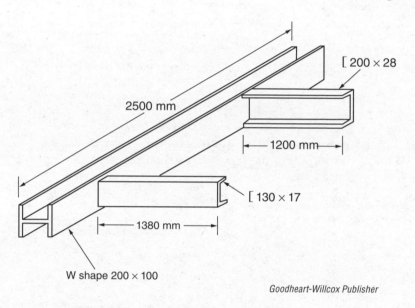

[200 × 28

2500 mm

1200 mm

[130 × 17

1380 mm

W shape 200 × 100

Goodheart-Willcox Publisher

15. A machine base was fabricated from the following list of materials. What is the weight of the machine base to the nearest pound?

Item	Quantity	Description	Size	Weight Per Foot
A	1	Steel plate	3/8″ × 25″ × 40″	—
B	2	Angle iron	2″ × 2″ × 1/8″ × 39″	1.65
C	4	Angle iron	2 1/2″ × 2″ × 3/8″ × 13 1/2″	5.3
D	6	Steel plate	1/2″ × 3″ × 3″	—
E	4	Angle iron	3″ × 3″ × 3/8″ × 17 5/8″	7.2
F	1	S shape	3″ × 5.7 lb × 2′–6″	5.7
G	11	Solid round bar	1 1/8″ dia. × 7″	3.38

Name _____ **Date** _____ **Class** _____

16. Calculate the weight of the following steel pillar to the nearest pound.

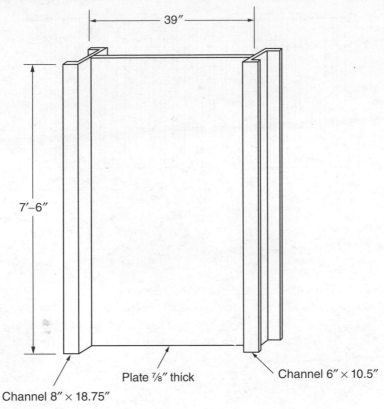

39″

7′–6″

Plate ⅞″ thick

Channel 6″ × 10.5″

Channel 8″ × 18.75″

Goodheart-Willcox Publisher

17. What is the weight of the following steel base for a swivel chair? Calculate your answer to the nearest gram.

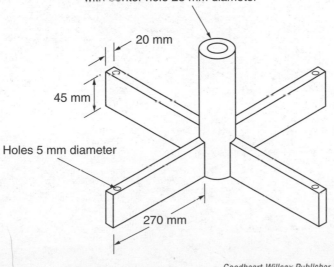

Hollow bar stock—length 175 mm × 60 mm diameter with center hole 28 mm diameter

20 mm

45 mm

Holes 5 mm diameter

270 mm

Goodheart Willcox Publisher

18. Calculate the weight of this section of steel fence to the nearest pound.

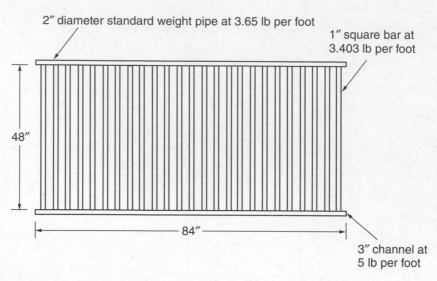

2" diameter standard weight pipe at 3.65 lb per foot

1" square bar at 3.403 lb per foot

48"

84"

3" channel at 5 lb per foot

Goodheart-Willcox Publisher

19. Calculate the weight of the following test piece to the nearest tenth of a pound. All plates are 3/8" thick and the overall length is 38". (Hint: In this type of problem, it is difficult to keep track of all the pieces. Try labeling the pieces as *a*, *b*, *c*, etc. Then, calculate the volume for each piece. Total all volumes and then calculate the total weight.)

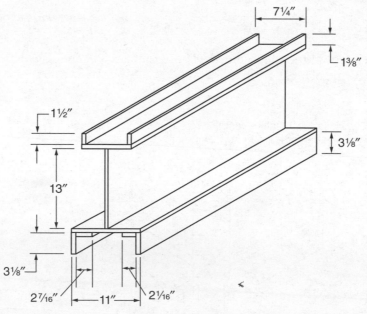

7¼"

1⅜"

1½"

3⅛"

13"

3⅛"

2⁷⁄₁₆"   11"   2¹⁄₁₆"

Goodheart-Willcox Publisher

Name _____   **Date** _____   **Class** _____

20. Calculate the weight of the following table frame to the nearest tenth
    of a pound.

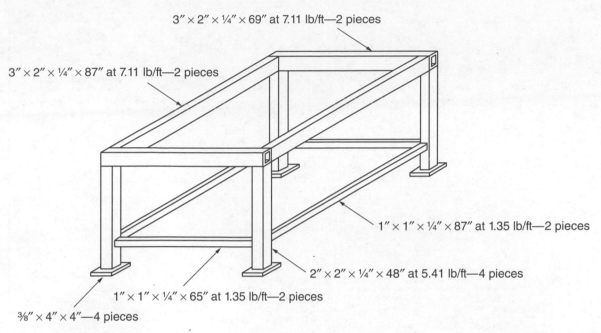

$3'' \times 2'' \times \frac{1}{4}'' \times 69''$ at 7.11 lb/ft—2 pieces

$3'' \times 2'' \times \frac{1}{4}'' \times 87''$ at 7.11 lb/ft—2 pieces

$1'' \times 1'' \times \frac{1}{4}'' \times 87''$ at 1.35 lb/ft—2 pieces

$2'' \times 2'' \times \frac{1}{4}'' \times 48''$ at 5.41 lb/ft—4 pieces

$1'' \times 1'' \times \frac{1}{4}'' \times 65''$ at 1.35 lb/ft—2 pieces

$\frac{3}{8}'' \times 4'' \times 4''$—4 pieces

Goodheart-Willcox Publisher

Work Space/Notes

UNIT 22

Bending Metal

Key Terms

flat-out length

Introduction

You will often be called on to bend metal as part of your job in a welding shop. When a piece of metal is permanently bent, its flat-out length is changed. *Flat-out length* is the original length of a piece of metal that has been permanently bent, shaped, or formed. Because of this dimensional change, you must be able to calculate flat-out lengths of metal objects that will be formed by bending. Most of the bends you will encounter can be grouped into the following types:

- 90° sharp corner bend
- Cylinder
- Half cylinder
- Quarter cylinder

A short explanation of the effects of bending may help you understand the calculations of bending. Why does permanent bending affect the length of a piece of metal? The primary reason is that metal can be stretched and compressed. For example, when metal is bent, the metal at the bend is compressed on the inside of the corner and stretched on the outside of the corner. The bent piece of metal now has two different lengths—the inside length and the outside length.

90° Sharp Corner Bend

A 90° sharp corner bend is the most common bend you will encounter. Drawings for bent metal parts can be dimensioned in two different ways. Either the inside lengths or the outside lengths can be dimensioned. The following illustrations show both methods.

When given the job of bending a piece of metal, you will need to know either the inside dimensions or the outside dimensions. You will also need to know the thickness of the metal. With this information, you can calculate the flat-out length.

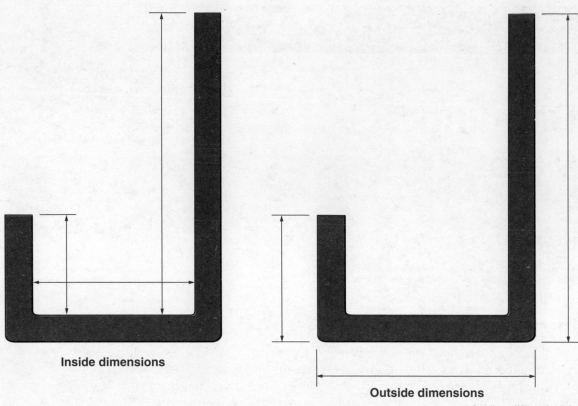

Inside dimensions

Outside dimensions

Goodheart-Willcox Publisher

Outside Dimensions

If you are given the outside dimensions, add all the outside lengths. Then subtract 1/2 the metal thickness for each 90° bend.

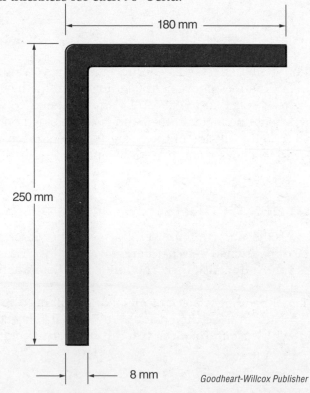

Goodheart-Willcox Publisher

180 mm + 250 mm = 430 mm – 4 mm = 426 mm
Flat-out length is 426 mm.

Inside Dimensions

If you are given the inside dimensions, add all the inside lengths. Then add 1/2 the metal thickness for each 90° bend.

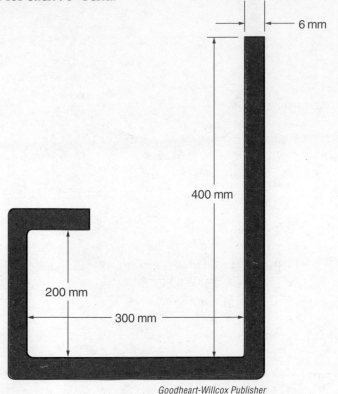

6 mm

400 mm

200 mm

300 mm

Goodheart-Willcox Publisher

200 mm + 300 mm + 400 mm = 900 mm + 3 mm + 3 mm + 3 mm = 909 mm
Flat-out length is 909 mm.

Cylinder

A piece of metal bent into a cylinder has two diameters—an inside diameter and an outside diameter. The difference between the inside diameter and the outside diameter is twice the thickness of the metal, as shown in the following illustration.

When given the job of bending a piece of metal into a cylinder, you will need to know either the inside or outside diameter, as well as the thickness of the metal. You must also know how to calculate the circumference of a circle (refer to Unit 19). With this information, you can calculate the flat-out length.

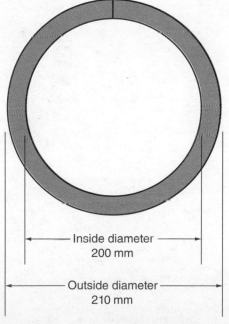

Inside diameter
200 mm

Outside diameter
210 mm

Goodheart-Willcox Publisher

Outside Diameter

To determine flat-out length when you are given the outside diameter, calculate the circumference using the outside diameter minus the metal thickness.

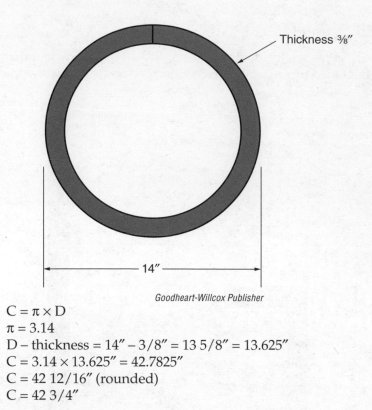

Thickness ⅜″

14″

Goodheart-Willcox Publisher

C = π × D
π = 3.14
D – thickness = 14″ – 3/8″ = 13 5/8″ = 13.625″
C = 3.14 × 13.625″ = 42.7825″
C = 42 12/16″ (rounded)
C = 42 3/4″

Flat-out length is 42 3/4″.

Inside Diameter

To determine flat-out length when you are given the inside diameter, calculate the circumference using the inside diameter plus the metal thickness.

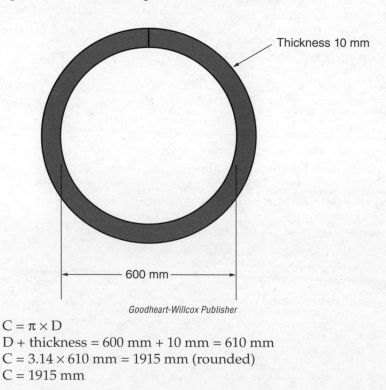

Thickness 10 mm

600 mm

Goodheart-Willcox Publisher

C = π × D
D + thickness = 600 mm + 10 mm = 610 mm
C = 3.14 × 610 mm = 1915 mm (rounded)
C = 1915 mm

Flat-out length is 1915 mm.

Half Cylinder

Half cylinder bends are calculated basically in the same way as full cylinder bends. The only difference is that a one-half circumference is calculated rather than a full circumference.

Outside Dimensions

To determine flat-out length of a half cylinder when you are given the outside dimensions, calculate the circumference using the outside diameter minus the metal thickness. Then divide the result in half. If the bent piece has legs that extend beyond the half cylinder, add the outside lengths to the one-half circumference to find the flat-out length.

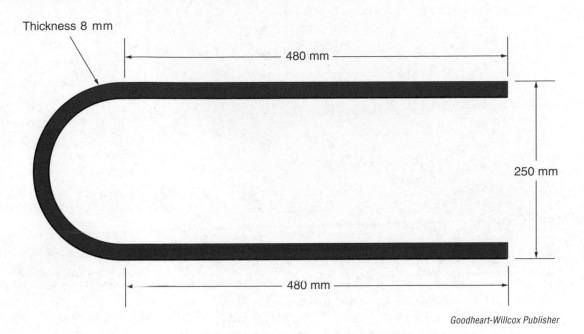

Goodheart-Willcox Publisher

$$C = \pi \times D$$
$$\pi = 3.14$$
$$D - \text{thickness} = 250 \text{ mm} - 8 \text{ mm} = 242 \text{ mm}$$
$$C = 3.14 \times 242 \text{ mm} = 759.88 \text{ mm}$$
$$\frac{C}{2} = 759.88 \ \frac{\text{mm}}{2} = 379.94 \text{ mm} = 380 \text{ mm (rounded)}$$
$$380 \text{ mm} + 480 \text{ mm} + 480 \text{ mm} = 1340 \text{ mm}$$

Flat out length is 1340 mm.

Inside Dimensions

To determine flat-out length of a half cylinder when you are given the inside dimensions, calculate the circumference using the inside diameter plus the metal thickness. Then divide the result in half. If the bent piece has legs that extend beyond the half cylinder, add the inside lengths to the one-half circumference to find the flat-out length.

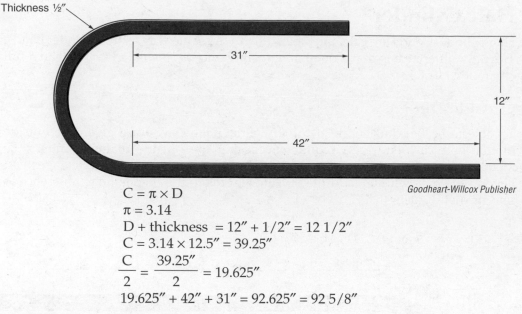

Thickness ½″

31″

42″

12″

Goodheart-Willcox Publisher

$C = \pi \times D$
$\pi = 3.14$
D + thickness = 12″ + 1/2″ = 12 1/2″
$C = 3.14 \times 12.5″ = 39.25″$
$$\frac{C}{2} = \frac{39.25″}{2} = 19.625″$$
19.625″ + 42″ + 31″ = 92.625″ = 92 5/8″

Flat-out length is 92 5/8″.

Quarter Cylinder

Quarter cylinder bends are calculated basically in the same way as full cylinder bends. The only difference is that one-quarter of a circumference is calculated rather than a full circumference.

Outside Dimensions

To determine flat-out length of a quarter cylinder when you are given the outside dimensions, calculate the circumference using the outside diameter minus the metal thickness. Then divide the result by four. If the bent piece has legs that extend beyond the quarter cylinder, add the outside lengths to the one-quarter circumference to find the flat-out length.

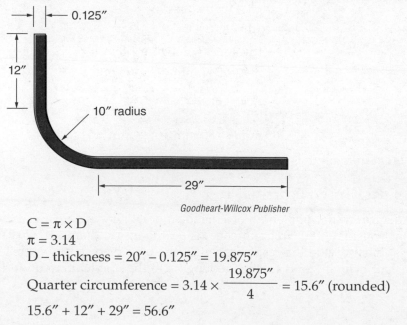

0.125″

12″

10″ radius

29″

Goodheart-Willcox Publisher

$C = \pi \times D$
$\pi = 3.14$
D – thickness = 20″ – 0.125″ = 19.875″
$$\text{Quarter circumference} = 3.14 \times \frac{19.875″}{4} = 15.6″ \text{ (rounded)}$$
15.6″ + 12″ + 29″ = 56.6″

Flat-out length is 56.6″.

Inside Dimensions

To determine flat-out length of a quarter cylinder when you are given the inside dimensions, calculate the circumference using the inside diameter plus the metal thickness. Then divide the result by four. If the bent piece has legs that extend beyond the quarter cylinder, add the inside lengths to the one-quarter circumference to find the flat-out length.

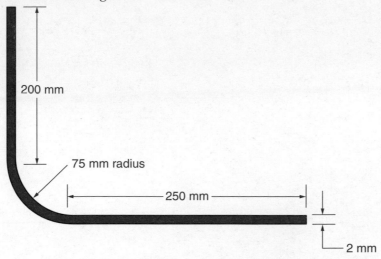

Goodheart-Willcox Publisher

$C = \pi \times D$

$\pi = 3.14$

D + thickness = 150 mm + 2 mm = 152 mm

Quarter circumference = $3.14 \times \dfrac{152 \text{ mm}}{4}$ = 119.32 mm = 119 mm (rounded)

119 mm + 200 mm + 250 mm = 569 mm

Flat-out length is 569 mm.

Work Space/Notes

Name _____ **Date** _____ **Class** _____

Perform the following equations as specified. Show all your work. Box your answers.

1. Seventeen brackets are required as shown. What is the total length of metal required?

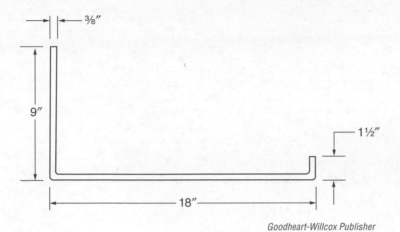

Goodheart-Willcox Publisher

2. Calculate the flat-out length.

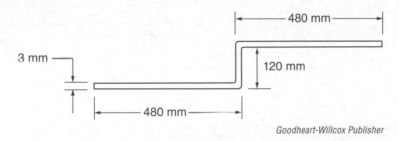

Goodheart-Willcox Publisher

3. Calculate the flat-out dimensions (length and width) of part Ⓐ.

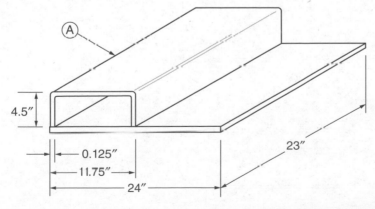

Goodheart-Willcox Publisher

4. Calculate the flat-out length of this 3/16" thick piece.

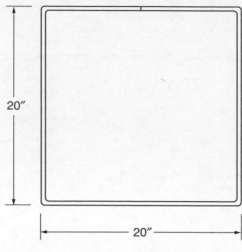

20"

20"

Goodheart-Willcox Publisher

5. Calculate the flat-out length of the piece in the following illustration.

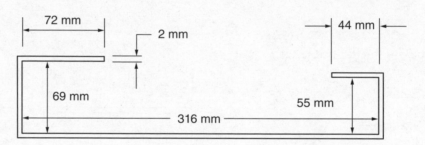

72 mm

2 mm

44 mm

69 mm

55 mm

316 mm

Goodheart-Willcox Publisher

6. What is the flat-out length and width of this 1/4" thick tank to the nearest thirty-second?

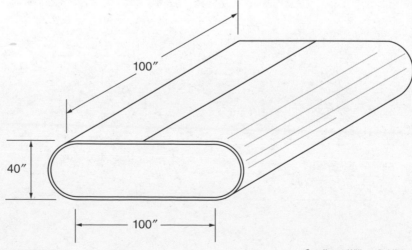

100"

40"

100"

Goodheart-Willcox Publisher

Name _____ **Date** _____ **Class** _____

7. Calculate the flat-out length of this 1/4″ thick ring to the nearest thirty-second.

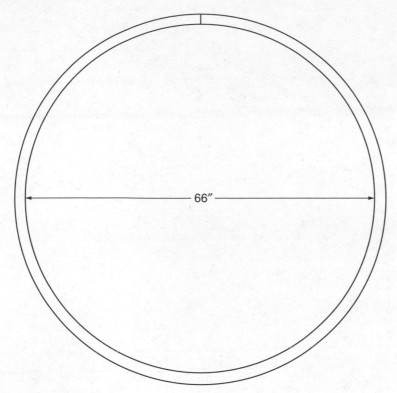

66″

Goodheart-Willcox Publisher

8. Calculate the flat-out length of this 3 mm thick piece.

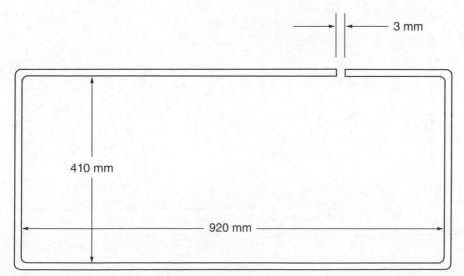

3 mm

410 mm

920 mm

Goodheart-Willcox Publisher

9. Calculate the flat-out dimension for this piece to the nearest sixty-fourth.

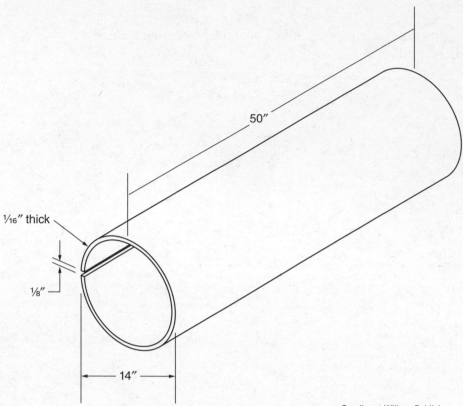

1/16″ thick

1/8″

14″

50″

Goodheart-Willcox Publisher

10. Calculate the flat-out length of this 2 mm thick piece of steel.

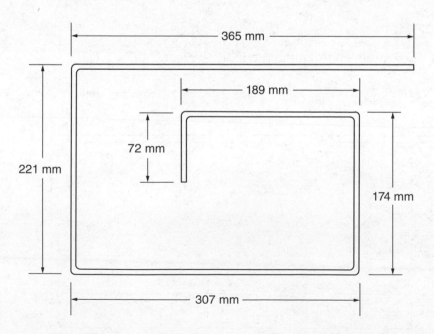

365 mm

189 mm

72 mm

221 mm

174 mm

307 mm

Goodheart-Willcox Publisher

Name _____   **Date** _____   **Class** _____

11. Calculate the flat-out length to the nearest sixty-fourth.

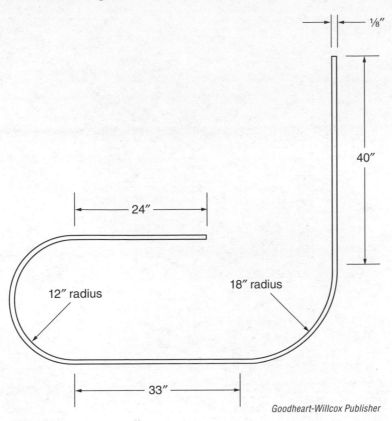

Goodheart-Willcox Publisher

12. Calculate the flat-out length of the following piece to the nearest millimeter.

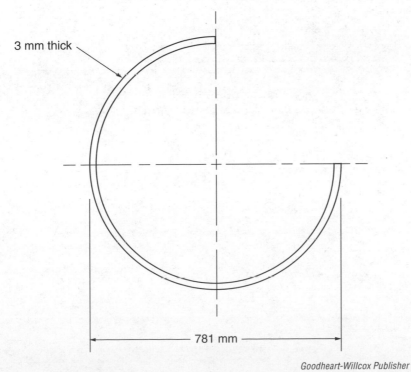

Goodheart-Willcox Publisher

13. Part Ⓐ and part Ⓑ are each 4 mm thick. Calculate the flat-out length of each part.

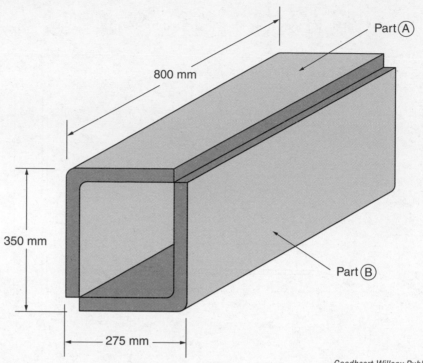

Goodheart-Willcox Publisher

14. Calculate the flat-out length.

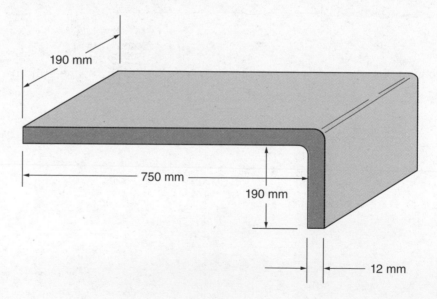

Goodheart-Willcox Publisher

Name _____   **Date** _____   **Class** _____

15. Calculate the flat-out length.

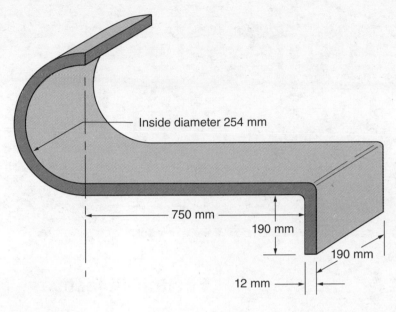

Inside diameter 254 mm

750 mm

190 mm

190 mm

12 mm

Goodheart-Willcox Publisher

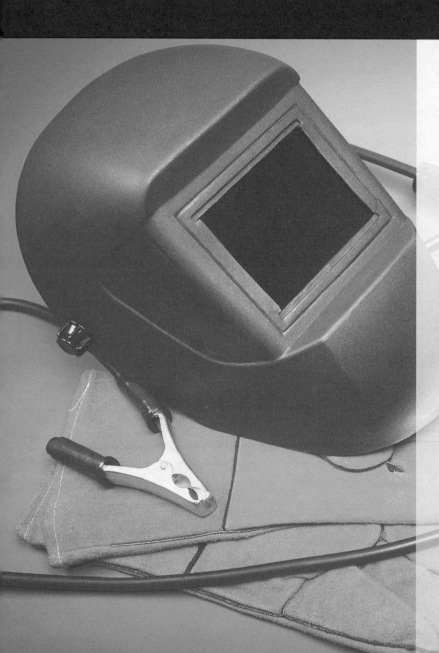

Percentages and the Metric System

Section Objectives

After studying this section, you will be able to:

- Define percentage.
- Convert between a fraction, a decimal, and a percentage.
- Calculate percentages.
- List the seven basic metric units.
- Explain the derived units of measure from length and mass.
- List the standard units of metric length.
- Convert measurements of length.
- Estimate and visualize metric lengths.
- Explain derived units of metric length.
- List the standard units of metric mass.
- Convert measurements of mass.
- Estimate and visualize metric mass.
- Explain derived units of metric mass.
- Properly use the language of SI.

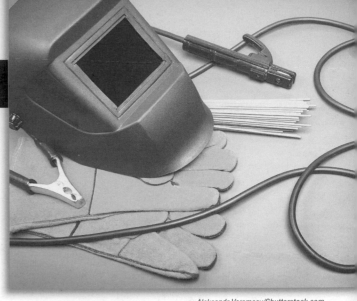

UNIT 23

Percentages

Key Terms

percent (%)

Introduction

Percent is a word that expresses things "by the hundred." Percents are parts of the whole of anything when it is divided into 100 equal parts. For example, 80 parts out of 100 can be expressed as 80%. The % symbol next to a number tells you it is a percentage number. The use of percents is common in daily working life. Items such as the following are usually expressed in percentages:

- The composition of steel
- The composition of alloys
- Salary deductions
- Scrap rate on a job
- Quality control statistics

Being able to solve problems involving percentage is a useful skill. Mathematically, a variety of things can be done with percent. Percents can be converted to decimals or fractions and vice versa. The percent of a number can be calculated. One number can be expressed as a percent of another.

Changing a Percent to a Fraction

Since percents refer to hundredths, changing them to fractions simply involves removing the percent sign and writing the number as a fraction with a denominator of 100. Reduce if necessary.

$$75\% = \frac{75}{100} = \frac{3}{4}$$

$$100\% = \frac{100}{100} = 1$$

$$125\% = \frac{125}{100} = 1\frac{25}{100} = 1\frac{1}{4}$$

$$3{,}000\% = \frac{3{,}000}{100} = 30$$

The above method works for most cases, but if the percent is a fraction or a mixed number, there is an additional step. First, change the mixed number to an improper fraction. Now that you have the percent in fractional form, remove the percent sign, multiply the denominator by 100, and then reduce, if necessary.

$$\frac{1}{4}\% = \frac{1}{400}$$

$$\frac{3}{16}\% = \frac{3}{1,600}$$

$$12\frac{1}{2}\% = \frac{25}{2}\% = \frac{25}{200} = \frac{1}{8}$$

$$107\frac{1}{2}\% = \frac{215}{2}\% = \frac{215}{200} = 1\frac{15}{200} = 1\frac{3}{40}$$

Changing a Percent to a Decimal

This is done using the same method as explained for changing percents to fractions, except the final answer is expressed as a decimal number. Remove the percent sign, divide the number by 100, and then express the answer as a decimal.

$$67\% = \frac{67}{100} = 0.67$$

$$16.6\% = \frac{16.6}{100} = 0.166$$

$$219\% = \frac{219}{100} = 2.19$$

$$1,019\% = \frac{1,019}{100} = 10.19$$

In cases where the percent is a fraction or a mixed number, first change the fraction to a decimal.

$$\frac{1}{16}\% = 0.0625\% = \frac{0.0625}{100} = 0.000625$$

$$11\frac{1}{8}\% = 11.125\% = \frac{11.125}{100} = 0.11125$$

$$95\frac{1}{2}\% = 95.5\% = \frac{95.5}{100} = 0.955$$

$$365\frac{1}{4}\% = 365.25\% = \frac{365.25}{100} = 3.6525$$

Changing a Fraction to a Percent

To easily change fractions, mixed numbers, and whole numbers to percents, multiply by 100 and add a percent sign.

$$\frac{9}{10} = \frac{9}{10} \times 100 = 90\%$$

$$1 = 1 \times 100 = 100\%$$

$$2\frac{3}{4} = \frac{11}{4} \times 100 = 275\%$$

$$5\frac{3}{7} = \frac{38}{7} \times 100 = 542\frac{6}{7}\%$$

Changing a Decimal to a Percent

To change decimals to percents, multiply by 100 and add a percent sign.

$$0.01 = 0.01 \times 100 = 1\%$$
$$0.25 = 0.25 \times 100 = 25\%$$
$$1.0 = 1.0 \times 100 = 100\%$$
$$3.5 = 3.5 \times 100 = 350\%$$

Calculating a Percent of a Number

This type of question is usually expressed as, "What is 25% of 300?" You may recall from Unit 9 that *of* means to multiply. Therefore, the solution is arrived at by multiplying 25% × the number in question. In this case, it is 300. To do this, change the percent to a decimal and proceed with the multiplication. In this example, 0.25 × 300 = 75. Twenty-five percent of 300 is 75.

$$30\% \text{ of } 3{,}750 = 0.30 \times 3{,}750 = 1{,}125$$
$$112\% \text{ of } 43 = 1.12 \times 43 = 48.16$$
$$1.5\% \text{ of } 900 = 0.015 \times 900 = 13.5$$

The percent can also be calculated in fractional form rather than as a decimal.

$$16\frac{2}{3}\% \text{ of } 2{,}700 = \frac{50}{3}\% \text{ of } 2{,}700 = \frac{50}{300} \times 2{,}700 = 450$$

Calculating the Percent One Number Is of Another Number

This type of question is usually expressed as, "What percent of 52 is 13?" The word *of* means *to multiply*, and *is* means *equals*. Therefore, the questions can be expressed in the following way.

$$?\% \times 52 = 13$$

To solve this problem, isolate the unknown percent value. This can be achieved by dividing both sides of the equation by 52.

$$\frac{?\% \times 52}{52} = \frac{13}{52}$$

On the left side of the equal symbol, the 52 below and the 52 above cancel out each other.

$$\frac{?\% \times \overset{1}{\cancel{52}}}{\underset{1}{\cancel{52}}} = \frac{13}{52}$$

This will form a fraction with the multiplier (52) as the denominator and 13 as the numerator. Since we are looking for a percent, we know that the denominator under the unknown value will be 100.

$$\frac{?}{100} = \frac{13}{52}$$

This equation can then be cross-multiplied.

$$\frac{?}{100} \diagdown \frac{13}{52}$$

We now need only to divide and plug the result back into the original statement—?% of 52 is 13. Remember that *of* means *to multiply* and *is* means *equals*.

$$\frac{1,300}{52} = 25$$

$$25\% \times 52 = 13$$

$$25\% \text{ of } 52 \text{ is } 13$$

The way in which this type of question is expressed may cause some confusion. For example, the question was originally expressed as "What percent of 52 is 13?" It could also have been expressed as, "Thirteen is what percent of 52?" Since *is* means equals and *of* means times, the resulting equation is $13 = ?\% \times 52$. Both expressions have the same meaning.

Name _____ **Date** _____ **Class** _____

Express the following percents as fractions. Show all your work. Be certain the columns line up. Box your answers.

1. 13%

2. 0.3%

3. 0.05%

4. $\dfrac{1}{16}$%

5. 1,010%

6. $\dfrac{2}{7}$%

Express the following percents as decimals. Show all your work. Be certain the columns line up. Box your answers.

7. 75%

8. 39%

9. 14.14%

10. 0.125%

11. 0.007%

12. $\dfrac{2}{5}$%

Express the following fractions as percents. Show all your work. Be certain the columns line up. Box your answers.

13. 3/4

14. 1/100

15. 331/3

16. 2,230 1/2

17. 11/19

18. 991/16

Express the following decimals as percents. Show all your work. Be certain the columns line up. Box your answers.

19. 0.125

20. 1.125

21. 0.019

22. 140.7

23. 0.003

24. 0.89

Calculate the following to two decimal places. Show all your work. Be certain the columns line up. Box your answers.

25. 35% of 823

26. 9.4% of 507

27. 100% of 28.5

28. $33\frac{1}{3}$% of 253

29. 0.17% of 123.6

30. 1,973% of 22

31. What percent of 423 is 95?

32. 579 is what percent of 606?

33. One (1) is what percent of 2?

34. What percent of 15 is 138?

35. What percent of 50 is 450?

36. 31.2 is what percent of 65.7?

37. A welder's suggestion to use a different type of fixture resulted in a 15% increase in production. The previous production rate was 180 parts per hour. What is the new production rate?

38. For year one, a group of mills produced 83 million tons of steel. For the second year, production increased by 8.5%. How many tons of steel were produced in the second year?

39. Three hundred and eighteen motorcycles started in the Illinois-Michigan Motorcycle Rally. Only eighty-five bikes finished the rally. What percent of the bikes did not finish? Calculate to the nearest percent.

Name _____   **Date** _____   **Class** _____

40. A welding shop was allowed a discount of 12 1/2% on the regular price of $426.00 for a pedestal grinder. Calculate the price the shop paid to the nearest cent.

41. The website for the Metal Building Association reported that sales this year were 32 1/2% ahead of last year's total of $1,266,000,000. What was the dollar value increase in metal building sales this year?

All of the employees listed in the following table received a salary increase of 4.75%. Calculate the weekly gross pay, income tax, union dues, and social security to the nearest cent and enter the amounts in the table.

Employees	Weekly Gross Pay		Income Tax		Union Dues	Social Security
	Before Increases	After Increases	%	$	3½% of gross pay	4% of gross pay
42. Calleja, Nick	$453.60		16%			
43. Ing, Richard	$491.60		18%			
44. Lima, Jose	$354.00		15%			
45. Massey, Vincent	$320.00		15%			
46. Shulman, Lorna	$450.80		19%			
47. Simpson, Katie	$402.00		18%			
48. Williams, Deroy	$458.80		15%			
49. Zakoor, Eli	$398.80		17%			

Goodheart-Willcox Publisher

Work Space/Notes

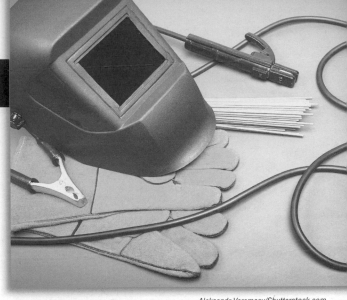

Aleksandr Veremeev/Shutterstock.com

The Metric System

Terms

derived units   metric ton

International System of Units (SI)

A Metric Workplace

Anyone who has been involved with the metalworking trades realizes that the global market is a powerful force. The decision by major manufacturers to employ the metric system has guaranteed its prominence in the workplace.

The welding industry is just one of many industries to adopt the metric system. As a welder, you will be expected to understand and work in metric. You will be pleased to know it is not a difficult system to learn. It is a logical, consistent system. However, the biggest difficulty most people have is changing established habits. Since many of us are accustomed to using the inch/lb units, we find that giving them up can be quite difficult. But, as job requirements include the ability to work in metric, an attachment to the inch/lb system usually gives way to the reality that an understanding of metrics is an important part of your training as a welder.

The Metric System

The metric system is a system of measurement that is used for everything measurable. The metric system started in the late eighteenth century in France. As time passed, new versions and variations were randomly added to the system, causing unnecessary complications. In 1960, after lengthy international discussions, the *International System of Units*, or *SI*, was established. This modern form of the metric system, also known as SI Metrics, is the most widely used measurement system in the world.

There are seven basic units of SI. The outstanding characteristic of the seven basic units is that any quantity in the universe can be measured by one of the seven basic units.

The Basic Units

The seven basic units of the metric system are as follows:

- Meter (m)—measuring length
- Kilogram (kg)—measuring mass
- Second (s)—measuring time
- Ampere (A)—measuring electrical current
- Kelvin (K)—measuring temperature
- Candela (cd)—measuring luminous intensity
- Mole (mol)—measuring the amount of substance

In welding, you will be required to know and apply the basic unit for measuring length and the basic unit for measuring mass. In everyday use, the word *weight* is often used in place of the word *mass*, although they have different meanings (see the Glossary). Be aware of the differences. However, in this text, we use the word *weight* as it is often understood in everyday speech, having the same meaning as the term *mass*.

Derived Units of Measure

Derived units of measure are additional units of measure in SI that are arrived at through equations using the basic units. See the following examples:

Derived Quantity	Basic Units	Derived Unit (symbol)
Area	$m \times m$	square meter (m^2)
Volume	$m \times m \times m$	cubic meter (m^3)

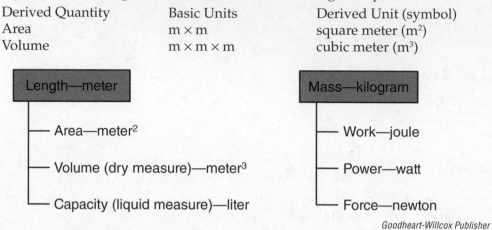

Goodheart-Willcox Publisher

It is usually not necessary for a welder to be completely familiar with the derived units of mass. However, the derived units of length are important to know.

Length

Length is one of the basic dimensions you will work with in the metric world. The originators of SI decided each unit of measure should have a base unit as a reference point. Then, subdivisions and multiples of the base unit were agreed on and named. Their lengths and names were all related to the base unit, making a consistent system. The meter was chosen as the base unit for length.

Standardized Units of Length

The following chart shows the base unit (the meter) and some of the subdivisions and multiples established. The very large multiples and the very small subdivisions have been omitted from the chart in order to focus on those most commonly used.

Name	Symbol	How Each Unit of Length Is Related to the Base Unit (Meter)
Kilometer	km	1000 meters
Hectometer	hm	100 meters
Dekameter	dam	10 meters
Meter	m	1 meter
Decimeter	dm	0.1 meter
Centimeter	cm	0.01 meter
Millimeter	mm	0.001 meter

Base unit →

Goodheart-Willcox Publisher

Start reading the chart at the base unit (the meter) and work your way up and down from there. Do not be concerned about what these lengths actually look like; you will become familiar with this later. For now, just observe how the system is organized and how the units are related. The names of all units end in *meter* and the abbreviations all end in the letter *m*. All units are related to the base unit in multiples of 10. The prefix indicates the number of meters in each unit. For example, *kilo* means 1,000, so a kilometer is 1000 meters.

In welding, you will not be working with the hectometer, dekameter, and decimeter. They are shown in the chart so you can see the regularity and pattern of the system.

Converting Metric Units of Length

The following diagram is a useful device for learning how to convert metric measurements of length.

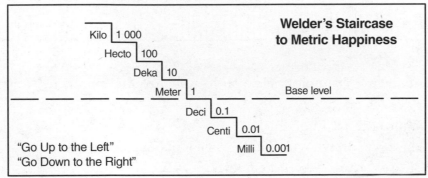

1. Convert 1629 m to kilometers.

   Looking at the stairway, you have to go up three steps (or places) to the left to move from meters to kilometers. So, in this case, you would move the decimal point three places to the left. Therefore, 1629 m equals 1.629 kilometers.

2. Convert 85 cm to millimeters.

   Looking at the stairway, you have to go down one step to the right to move from centimeters to millimeters. In this case, you would move the decimal point one place to the right. Therefore, 85 cm equals 850 mm.

3. Convert 7.5 km to decimeters.

   Go down the stairway four places to the right. Move the decimal point four places to the right. Therefore, 7.5 km equals 75000 dm.

The "metric stairway" is a useful visual tool for converting measurements. There is, of course, a mathematical explanation why conversion from one unit to another can be done simply by moving the decimal point. Since all units are related in multiples of ten, converting consists of multiplying by these multiples and, in effect, moving the decimal.

4. Convert 5.9 km to meters.

   $5.9 \text{ km} \times 1000 = 5900 \text{ m}$

5. Convert 16 cm to millimeters.

   $16 \text{ cm} \times 10 = 160 \text{ mm}$

6. Convert 1437 dm to hectometers.

   $1437 \text{ dm} \times 0.001 = 1.437 \text{ hm}$

Visualizing Metric Lengths

Becoming knowledgeable in the metric system requires that you develop an awareness of the actual lengths of metric units. Just as you already have a rough idea of the length of one inch, one foot, and one yard, you can also start becoming aware of metric lengths. In order to avoid being overwhelmed with all the units of length, it is best to concentrate just on the millimeter, centimeter, and meter. Being familiar with these three will satisfy your needs in welding. In fact, almost all the metric blueprints you will be reading are dimensioned only in millimeters. Below is a full size metric scale, 150 mm long. If you do not own a metric scale, you should buy one now.

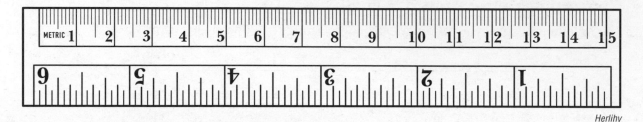

Herlihy

The following examples help you to visualize the millimeter, centimeter, and meter.

- A dime is about 1 mm thick.
- A thick line drawn with a dull pencil is about 1 mm wide.
- A slice of bread is about 1 cm thick.
- A stack of ten dimes is about 1 cm high.
- One meter is about three inches longer than one yard.
- The height of a doorknob is about one meter from the floor.

Estimating Metric Lengths

The following useful technique involves memorizing certain body dimensions and then using these to estimate other dimensions.

Measure the following to the nearest millimeter (mm) and record the results.

The width of your thumb in millimeters:

The width of your palm in millimeters:

The span of your hand from the smallest finger to thumb in centimeters:

The distance from your elbow to your fingertips in centimeters:

The length of your stride in centimeters:

Your height in meters:

In the following exercise, you will estimate the length of your welding booth using the width of your hand as a guide.

1. Count the number of hand widths it takes to span the length of the booth.
2. Multiply the number of hand widths by the width of your hand to arrive at an estimate. For example, if your hand has a width of 100 mm, and it takes 13 hand widths to span the booth, then the estimated length is 13×100, or 1300 mm.
3. Check your accuracy with a metric scale.

Converting Metric and US Customary Units of Length

At some time in your job, you will probably have to make conversions between these two systems. Conversion charts or cards are often available, but you should know the mathematical method for converting. Your main concern as a welder will be converting from millimeters to inches and from inches to millimeters. Therefore, these are the only conversions that will be explained here.

Calculations in converting often require rounding the answer. Answers can be rounded for maximum mathematical accuracy or for the accuracy normally required in the shop. A method for shop accurate conversions is explained in the following section.

Shop Acceptable Accuracy

In general, the closest tolerance to which a welder is expected to work is to the nearest 1/64″. The following guidelines ensure that your work stays within this degree of accuracy.

1. When converting from inches to millimeters, round off to the nearest millimeter.
2. When converting from millimeters to inches, round off to the nearest 0.1″. Study the following examples.

Convert 7.367″ to mm.
Exact conversion: $7.367 \times 25.4 = 187.1218$ mm
Shop acceptable accuracy = 187 mm

Convert 11 1/16″ to mm.
Exact conversion: $11.0625 \times 25.4 = 280.9875$ mm
Shop acceptable accuracy = 281 mm

Convert 65.5 mm to inches.
Exact conversion: $65.5 \div 25.4 = 2.5787402$″
Shop acceptable accuracy = 2.6″

Convert 485.35 mm to inches.
Exact conversion: $485.35 \div 25.4 = 19.108268$″
Shop acceptable accuracy = 19.1″

Two rules for converting are as follows:

- When converting inches to millimeters, multiply by 25.4.
- When converting millimeters to inches, divide by 25.4.

Derived Units of Length

As mentioned earlier, the basic unit of length has three derived units of length used to measure area, volume, and capacity.

Area

Area is calculated in the normal manner, depending on the shape of the object. Results are expressed as mm^2, km^2, etc. Be sure you are calculating with the same units; that is, $mm \times mm$ or $cm \times cm$. For example, do not multiply meters times millimeters to calculate an area.

Volume

Volume refers to dry measure and is calculated in the normal manner, depending on the shape of the object. Results are expressed as m^3, mm^3, etc. As pointed out with area, be certain you are calculating with the same units—$m \times m \times m = m^3$.

Capacity

The liter is the base unit used for liquid measure, or capacity. The standardized units of capacity are illustrated in the following chart.

Name	Symbol	How Each Unit of Capacity Is Related to the Base Unit (Liter)
Kiloliter	kl	1000 liters
Hectoliter	hl	100 liters
Dekaliter	dal	10 liters
Liter	L	1 liter
Deciliter	dl	0.1 liters
Centiliter	cl	0.01 liters
Milliliter	ml	0.001 liters

Base unit → (points to Liter row)

Goodheart-Willcox Publisher

Similar to units of length, there is a regular pattern of units of capacity. Also, as with units of length, you will work with only a few of the units of capacity, such as the liter (L) and milliliter (ml). The other units are shown to illustrate the pattern of the system.

Converting Metric Units of Capacity

To convert from one unit of measure to another, use the following diagram. Also, study the examples that follow.

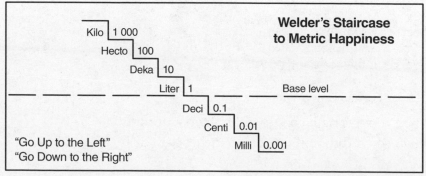

Goodheart-Willcox Publisher

Convert 49 liters (L) to milliliters (ml).
49 L = 49000 ml

Convert 7 kiloliters (kl) to liters.
7 kl = 7000 L

Convert 28 milliliters to liters.
28 ml = 0.028 L

Visualizing Metric Capacity

The following examples will help you visualize the liter and milliliter:

- A teaspoon is about 5 ml.
- A bowl of soup is about 200 ml.
- The liter is approximately the volume of four paper coffee cups.
- A tea kettle has a capacity of about 2 L.

Converting Metric and US Customary Units of Capacity

To convert between metric and standard capacity, use the conversion figure of one gallon = 3.785 liters.

Convert 27 gallons to liters (L).
27 gallons = 27 × 3.785 = 102.195 liters

Convert 11775 liters to gallons.
11775 liters = 11775 ÷ 3.785 = 3110.9643 gallons

Follow these rules for converting between metric and US customary units of capacity:

- When converting from gallons to liters, multiply by 3.785.
- When converting from liters to gallons, divide by 3.785.

Mass

You will often use the metric system to measure mass. The base unit for mass is the gram. The following chart shows the base unit and some of the subdivisions and multiples established for metric mass. The very large multiples and the very small subdivisions have been omitted in order to focus on those that are most commonly used.

Name	Symbol	How Each Unit of Mass Is Related to the Base Unit (Gram)
Megagram	Mg (t)	1000000 grams
—	—	—
—	—	—
Kilogram	kg	1000 grams
Hectogram	hg	100 grams
Dekagram	dag	10 grams
Gram	g	1 gram
Decigram	dg	0.1 gram
Centigram	cg	0.01 gram
Milligram	mg	0.001 gram

Base unit ———→ Gram

Goodheart-Willcox Publisher

Start reading the chart at the base unit and work your way up and down from there. Observe how the system is organized. All units are related to the base unit in multiples of 10. The names of all units end in *gram*, and the abbreviations all end in the letter *g*. There is one variation accepted here. The megagram is also commonly referred to as a **metric ton** and has the symbol *t*.

Of the eight units of mass shown on the chart, you will probably encounter only the megagram (metric ton), kilogram, gram, and milligram. The other units are illustrated so you can see the pattern of the system. Since the units increase in multiples of 10, you may have expected a metric unit for 10000 grams and 100000 grams. However, there are no metric units between the kilogram (1000 grams) and the metric ton (1000000 grams).

Converting Metric Units of Mass

The following diagram is a useful device for learning how to convert measurements. Review it before reading through the examples below.

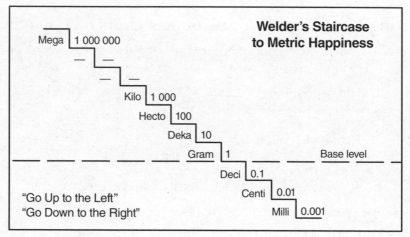

Goodheart-Willcox Publisher

Convert 3275 g to kilograms (kg).
You have to go up three steps (or places) to the left to move from gram to kilograms. So, you would move the decimal point three places to the left. Therefore, 3275 g equals 3.275 kg.

Convert 987000 milligrams (mg) to kilograms.
You have to go up six places to the left to move from milligrams to kilograms. Move the decimal point six places to the left. Therefore, 987000 mg equals 0.987 kg.

Convert 48 megagrams (Mg) to kilograms.
You have to move down three places to the right to move from megagrams to kilograms. Move the decimal point three places to the right. Therefore, 48 Mg equals 48000 kg.

As you can see from these and previous examples, converting between units in metric can be done by simply moving the decimal point.

Visualizing Metric Mass

The metric weights you should become familiar with are gram, kilogram, and metric ton. The following are some examples:

- A paper clip has a mass of about one gram.
- A thumbtack weighs about one gram.
- The mass of about five average sized apples is about one kilogram.
- A heavyweight boxer weighs about 91 kilograms.
- A compact car has a mass of about one metric ton.
- One metric ton weighs about 200 lb more than one standard ton.

Converting Metric and US Customary Units of Mass

$$1 \text{ lb} = 0.4536 \text{ kg}$$
$$1 \text{ kg} = 2.205 \text{ lb}$$
$$1 \text{ metric ton (t)} = 2,205 \text{ lb}$$

The above conversions have been rounded and, therefore, slight inaccuracies will result. They are accurate enough, however, for the work you will encounter in the shop. Below are some examples.

Convert 875 lb to kilograms.
$875 \times 0.4536 = 396.9$ lb

Convert 32,000 lb to metric tons.
$32,000 \div 2,205 = 14.5$ metric tons (rounded)

Convert 81.8 kg to lb.
$81.8 \times 2.205 = 180.4$ lb (rounded)

Convert 100000 kg to standard tons.
$$\frac{100,000 \times 2.205}{2,000} = 110.25 \text{ tons}$$

Convert 5 metric tons to lb.
$5 \times 2205 = 11,025$ lb

Convert 30 metric tons to standard tons.
$$\frac{30 \times 2,205}{2,000} = 33 \text{ standard tons (rounded)}$$

The Language of SI

1. Never use a period after a symbol unless it is at the end of a sentence.

   Example: Her height is 148 cm, but his height is 185 cm.

2. Symbols are never made plural by adding an *s*.

   Example: 1 g   10 g   100 g   1000 g

3. Symbols are almost always written in lowercase. The two exceptions so far are megagram and liter.

   Example: m for meter, g for gram, but Mg for megagram and L for liter.

4. Never start a sentence with a symbol. Write the name out in full.

   Example: Kilometers measure distance.

5. Always leave a space between the quantity and the symbol.

   Example: 15 km 6 m 24 g

6. If there is no quantity with the unit, spell the unit out in full.

   Example: Meters are marvelous, and grams are great.

7. Always use decimals, not fractions.

   Example: 0.25 g 1.5 2.75 m

8. Always place a zero before the decimal point if the value is less than one.

   Example: 0.125 L 0.75 kg 0.5 m

Name _____ **Date** _____ **Class** _____

1. List the seven basic units in the metric system.

2. Which two of the seven basic units will welders encounter most often?

3. Name the base unit for metric length.

4. How many millimeters are there in one meter?

5. How many centimeters are there in one meter?

Convert the following measurements. Show all your work. Box your answers.

6. 3.5 kilometers to millimeters

7. 1597 centimeters to meters

8. 12800 millimeters to kilometers

9. 2985400 meters to kilometers

10. 1000 centimeters to millimeters

11. 595 decimeters to kilometers

Convert the following measurements to the nearest millimeter. Show all your work. Box your answers.

12. 0.375″ to millimeters

13. 36″ to millimeters

14. 5′-8 1/2″ to millimeters

15. 16 1/2″ to millimeters

16. 100′ to meters

17. 100 yd to meters

Convert the following measurements to the nearest 1/64 inch. Show all your work. Box your answers.

18. 1000 mm to inches

19. 87.6 mm to inches

20. 112.8 mm to inches

21. 1000 m to feet and inches

22. 3855 mm to feet and inches

23. 10 km to feet and inches

24. Name the base unit for liquid capacity.

25. How many milliliters are there in a liter?

Convert the following. Show all your work. Box your answers.

26. 7570 L to gallons

27. 182 ml to liters

28. 54 gal to liters

29. 6.7 L to milliliters

30. 1 gal to milliliters

31. 10000 ml to gallons (to nearest hundredth)

32. Name the base unit for mass.

33. How many grams are there in one kilogram?

Convert the following. Show all your work. Box your answers.

34. 76950 g to milligrams

35. 39.25 kg to grams

36. 0.78 kg to milligrams

37. 212 mg to grams

38. 1000 mg to kilograms

39. 65 g to kilograms

Convert the following. Show all your work. Box your answers.

40. 1,000 lb to kilograms

41. 6472 g to lb (round to nearest hundredth)

42. 105 lb to grams

43. 3.5 standard tons to kilograms (round to nearest kilogram)

44. 4.25 t to standard tons (round to nearest hundredth)

45. 69 standard tons to kilograms (round to nearest kilogram)

Appendix

Useful Information

The following pages include several charts and tables filled with information that will prove useful to you throughout your career as a welder. This information includes decimal and fractional equivalents, conversions between US Customary and metric measurements, steps and calculations for bending metal, and a summary of formulas. Become familiar with this useful information.

Fraction, Decimal, and Millimeter Conversions

Fractional Inch	Decimal Inch	Millimeter	Fractional Inch	Decimal Inch	Millimeter
1/64	.015625	0.397	33/64	.515625	13.097
1/32	.03125	0.794	17/32	.53125	13.494
3/64	.046875	1.191	35/64	.546875	13.891
1/16	.0625	1.588	9/16	.5625	14.288
5/64	.078125	1.984	37/64	.578125	14.684
3/32	.09375	2.381	19/32	.59375	15.081
7/64	.109375	2.778	39/64	.609375	15.478
1/8	.125	3.175	5/8	.625	15.875
9/64	.140625	3.572	41/64	.640625	16.272
5/32	.15625	3.969	21/32	.65625	16.669
11/64	.171875	4.366	43/64	.671875	17.066
3/16	.1875	4.762	11/16	.6875	17.462
13/64	.203125	5.159	45/64	.703125	17.859
7/32	.21875	5.556	23/32	.71875	18.256
15/64	.234375	5.953	47/64	.734375	18.653
1/4	.25	6.350	3/4	.75	19.05
17/64	.265625	6.747	49/64	.765625	19.447
9/32	.28125	7.144	25/32	.78125	19.844
19/64	.296875	7.541	51/64	.796875	20.241
5/16	.3125	7.938	13/16	.8125	20.638
21/64	.328125	8.334	53/64	.828125	21.034
11/32	.34375	8.731	27/32	.84375	21.431
23/64	.359375	9.128	55/64	.859375	21.828
3/8	.375	9.525	7/8	.875	22.225
25/64	.390625	9.922	57/64	.890625	22.622
13/32	.40625	10.319	29/32	.90625	23.019
27/64	.421875	10.716	59/64	.921875	23.416
7/16	.4375	11.112	15/16	.9375	23.812
29/64	.453125	11.509	61/64	.953125	24.209
15/32	.46875	11.906	31/32	.96875	24.606
31/64	.484375	12.303	63/64	.984375	25.003
1/2	.5	12.700	1	1.0	25.400

Goodheart-Willcox Publisher

Useful Conversions between the SI Metric and US Customary Systems

Length	1 inch	= 25.4 mm	1 mm	= 0.03937 inches
	1 inch	= 2.54 cm	1 cm	= 0.3937 inches
	1 foot	= 30.48 cm	1 cm	= 0.0328 feet
	1 foot	= 0.3048 m	1 m	= 3.28 feet
	1 yard	= 0.9144 m	1 m	= 1.0936 yard
	1 mile	= 1.609 km	1 km	= 0.621 mile
Area	1 in^2	= 645.2 mm^2	1 mm^2	= 0.00155 in^2
	1 in^2	= 6.452 cm^2	1 cm^2	= 0.155 in^2
	1 ft^2	= 929.03 cm^2	1 cm^2	= 0.00108 ft^2
	1 ft^2	= 0.093 m^2	1 m^2	= 10.764 ft^2
	1 yd^2	= 0.836 m^2	1 m^2	= 1.196 yd^2
	1 square mile	= 2.59 km^2	1 km^2	= 0.386 square mile
Volume (dry)	1 in^3	= 16.388 cm^3	1 cm^3	= 0.061 in^3
	1 ft^3	= 0.028 m^3	1 m^3	= 35.31 ft^3
	1 yd^3	= 0.765 m^3	1 m^3	= 1.308 yd^3
Capacity (fluid)		US		Canadian
	1 quart	= 0.946 L	1.14 L	
	1 gallon	= 3.785 L	4.546 L	
	1 L	= 1.057 quarts	0.877 quarts	
	1 L	= 0.264 gallons	0.220 gallons	
Weight	1 oz	= 28.35 g	1 g	= 0.0353 oz
	1 lb	= 453.59 g	1 kg	= 2.205 lb
	1 lb	= 0.4536 kg	1 ton (metric)	= 2,204.6 lb

Goodheart-Willcox Publisher

Note: Converting between the US customary and metric systems often results in highly unwieldy numbers. Because of this, many of the above conversions have been rounded, but are quite accurate as a quick reference.

Bending Metal
(How to Calculate Flat Stock Lengths Assuming a Zero Bend Radius)

90° Corner Bend	Cylinder
1. Given the outside dimensions, add all the outside lengths. Then subtract 1/2 the metal thickness for each 90° bend. 2. Given the inside dimensions, add all the inside lengths. Then add 1/2 the metal thickness for each 90° bend.	1. Given the outside diameter, calculate the circumference using the outside diameter minus the metal thickness. 2. Given the inside diameter, calculate the circumference using the inside diameter plus the metal thickness.
Half Cylinder	**Quarter Cylinder**
1. Given the outside dimensions, calculate the circumference using the outside diameter minus the metal thickness. Then divide the result in half. 2. Given the inside dimensions, calculate the circumference using the inside diameter plus the metal thickness. Then divide the result in half.	1. Given the outside dimensions, calculate the circumference using the outside diameter minus the metal thickness. Then divide the result by four. 2. Given the inside dimensions, calculate the circumference using the inside diameter plus the metal thickness. Then divide the result by four.

Goodheart-Willcox Publisher

Inches Converted to Decimals of Feet

Inches	Decimal of a Foot	Inches	Decimal of a Foot	Inches	Decimal of a Foot
1/8	.01042	3 1/8	.26042	6 1/4	.52083
1/4	.02083	3 1/4	.27083	6 1/2	.54167
3/8	.03125	3 3/8	.28125	6 3/4	.56250
1/2	.04167	3 1/2	.29167	7	.58333
5/8	.05208	3 5/8	.30208	7 1/4	.60417
3/4	.06250	3 3/4	.31250	7 1/2	.62500
7/8	.07291	3 7/8	.32292	7 3/4	.64583
1	.08333	4	.33333	8	.66666
1 1/8	.09375	4 1/8	.34375	8 1/4	.68750
1 1/4	.10417	4 1/4	.35417	8 1/2	.70833
1 3/8	.11458	4 3/8	.36458	8 3/4	.72917
1 1/2	.12500	4 1/2	.37500	9	.75000
1 5/8	.13542	4 5/8	.38542	9 1/4	.77083
1 3/4	.14583	4 3/4	.39583	9 1/2	.79167
1 7/8	.15625	4 7/8	.40625	9 3/4	.81250
2	.16666	5	.41667	10	.83333
2 1/8	.17708	5 1/8	.42708	10 1/4	.85417
2 1/4	.18750	5 1/4	.43750	10 1/2	.87500
2 3/8	.19792	5 3/8	.44792	10 3/4	.89583
2 1/2	.20833	5 1/2	.45833	11	.91667
2 5/8	.21875	5 5/8	.46875	11 1/4	.93750
2 3/4	.22917	5 3/4	.47917	11 1/2	.95833
2 7/8	.23959	5 7/8	.48958	11 3/4	.97917
3	.25000	6	.50000	12	1.00000

Goodheart-Willcox Publisher

Summary of Formulas

Perimeter of a Square

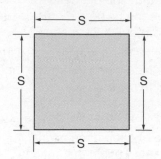

P = Perimeter
S = Side
P = S + S + S + S

Area of a Rectangle

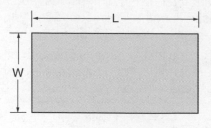

A = Area
L = Length
W = Width
A = L × W

Area of a Square

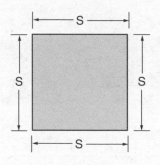

A = Area
S = Side
A = S × S

Perimeter of a Parallelogram

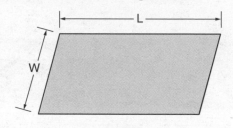

P = Perimeter
L = Length
W = Width
P = L + L + W + W

Perimeter of a Rectangle

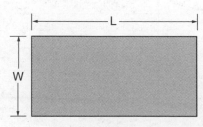

P = Perimeter
L = Length
W = Width
P = L + L + W + W

Area of a Parallelogram

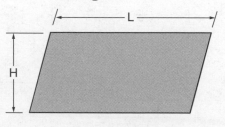

A = Area
L = Length
H = Height
A = L × H

Perimeter of a Trapezoid

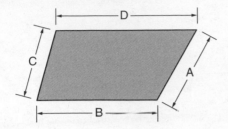

P = Perimeter
A = Length of a Side
B = Length of a Side
C = Length of a Side
D = Length of a Side
P = A + B + C + D

Perimeter of a Triangle

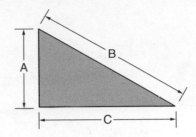

P = Perimeter
A = Length of a Side
B = Length of a Side
C = Length of a Side
P = A + B + C

Area of a Trapezoid

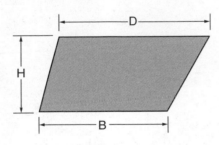

A = Area
B = Length of a Parallel Side
D = Length of a Parallel Side
H = Height

$$A = \frac{1}{2} \times (B + D) \times H$$

Note: To use this formula, first add the two values inside the parentheses, then perform the multiplication.

Area of a Triangle

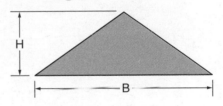

A = Area
B = Length of the Base
H = Height

$$A = \frac{1}{2} \times B \times H$$

Circumference of a Circle

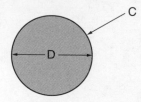

C = Circumference

D = Diameter

$\pi = 3.14$

$C = \pi \times D$

Diameter of a Circle

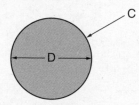

D = Diameter

C = Circumference

$\pi = 3.14$

$D = \dfrac{C}{\pi}$

Area of a Circle

A = Area

R = Radius

$\pi = 3.14$

$A = \pi \times R \times R$

Perimeter of a Half Circle

P = Perimeter

D = Diameter

$\pi = 3.14$

$P = \left(\dfrac{\pi \times D}{2}\right) + D$

Note: To use this formula, first perform the calculations inside the parentheses, then add the result to D.

Area of a Half Circle

A = Area

R = Radius

$\pi = 3.14$

$A = \dfrac{\pi \times R \times R}{2}$

Note: To use this formula, first perform the multiplication, then divide the result by 2.

Area of the Curved Surface of a Cylinder

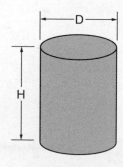

A = Area

D = Diameter

H = Height

$\pi = 3.14$

$A = \pi \times D \times H$

Perimeter of a Semicircular Sided Shape

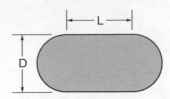

P = Perimeter

D = Diameter

L = Length of Straight Side

$$P = (\pi \times D) + (2 \times L)$$

Note: To use this formula, first perform the multiplication inside the parentheses, then add the two results together.

Area of a Semicircular Sided Shape

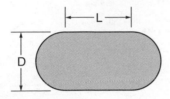

A = Area

D = Diameter

L = Length of Straight Side

$$A = \left(\frac{\pi \times D \times D}{4}\right) + (L \times D)$$

Note: To use this formula, first calculate each of the parts in parentheses, then add the two results together.

Volume of a Cube

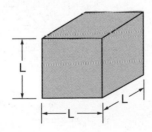

V = Volume

L = Length of a Side

$$V = L \times L \times L$$

Volume of a Rectangular Solid

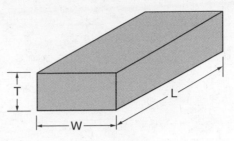

V = Volume

T = Thickness

W = Width

L = Length

$$V = T \times W \times L$$

Volume of a Solid Parallelogram

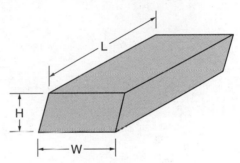

V = Volume

H = Height

W = Width

L = Length

$$V = H \times W \times L$$

Volume of a Solid Trapezoid

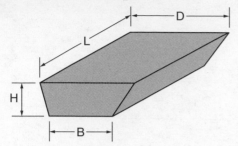

V = Volume

H = Height

B = Length of a Parallel Side

D = Length of a Parallel Side

$$V = \frac{1}{2} \times (B + D) \times H \times L$$

Note: To use this formula, first add the two values inside the parentheses, then perform the multiplication.

Volume of a Solid Triangle

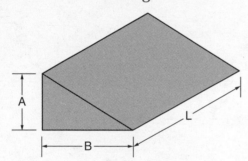

V = Volume

A = Length of a Side

B = Length of a Side

L = Length of a Side

$$V = \frac{1}{2} \times A \times B \times L$$

Volume of a Cylinder

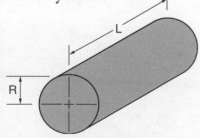

V = Volume

R = Radius

L = Length

π = 3.14

$$V = \pi \times R \times R \times L$$

Volume of a Solid Half Circle

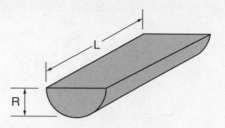

V = Volume

R = Radius

L = Length

π = 3.14

$$V = \frac{\pi \times R \times R \times L}{2}$$

Note: To use this formula, first perform the multiplication, then divide the result by 2.

Volume of a Solid Semicircular Sided Shape

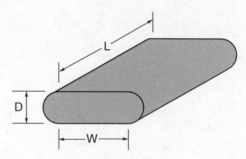

V = Volume

D = Diameter

W = Length of Straight Side

L = Length

π = 3.14

$$V = \left(\frac{\pi \times D \times D \times L}{4} \right) + (D \times W \times L)$$

Note: To use this formula, first perform the calculations inside the parentheses, then add the two results together.

Glossary

A

acetylene gas: A gas used in combination with oxygen as a fuel for welding and cutting. (13)

addition: The process of combining two or more individual numbers to form a single number that usually has a higher value than any of the previously individual numbers. (2)

alloy: An *alloy* is a blend of two or more elements, usually metals. Alloys are made to achieve certain special characteristics that do not occur in pure metals. (23)

aluminum flux: A flux designed for brazing aluminum. See *flux*. (5)

angle: The opening between two intersecting lines. (16)

angular measure: Refers to measuring the angle formed by two intersecting lines. (15)

approximate numbers: Numbers that are not perfectly accurate. For example, the number 16,297 can be written as the approximate number 16,300. (1)

Arabic number system: The number system developed in the Arab culture that is now in common usage. The system is made up of the ten digits 0, 1, 2, 3, 4, 5, 6, 7, 8, 9. Also referred to as *decimal number system*. (1)

area: The entire surface measure of a shape. (1, 17)

B

basic operations: The mathematical operations of addition, subtraction, multiplication, and division. (2)

borrow: A step in subtraction and division processes in which a number from the place value to the left is used to create a new number. (3)

brass: An alloy made up of copper and zinc. (7)

C

C shape: An abbreviation for *standard channel*. A structural steel product with a "[" cross section. (21)

carrying: A step in the addition and multiplication processes of whole numbers and decimal numbers in which the calculated value of a number combination is more than one single digit and the higher valued digits are added to the calculated value of the numbers in the column to the left. (2, 4)

cast iron: Cast iron is an alloy of iron and carbon. To be classified as cast iron, the carbon content must be more than 1.7%. (12, 20)

centimeter (cm): A metric measurement of length. It is equal to 10 millimeters, 0.01 meter, or approximately 2.54". (21)

circular measure: Refers to measuring curved lines. (15)

circumference: The perimeter of a circle. (19)

cold galvanizing compound: A product that can be applied to steel for the purpose of protecting it from corrosion. See *galvanizing*. (14)

cold-rolled steel: A process used in making steel. Cold-rolled steel is produced in its final form by passing steel between successive rollers. Both the steel and the rollers are cold, which results in a very accurate steel product with a smooth, shiny surface. Compare with *hot-rolled steel*. (3)

common denominator: The same denominator found in two or more fractions. For example, both 3/15 and 2/15 have a denominator of 15. (7)

common fraction: A fractional part expressed with a numerator and a denominator. (6)

complex fraction: A fraction in which the numerator, the denominator, or both contain a fraction. (10)

The number in parentheses following each definition indicates the chapter in which the term can be found.

D

decimal number: A number that is expressed with a decimal point. Numbers to the left of the decimal point indicate whole numbers. Numbers to the right of the decimal point indicate a value less than 1. For example, 1.3 or 0.5. (11)

decimal number system: The number system in common usage having the ten digits 0, 1, 2, 3, 4, 5, 6, 7, 8, 9. Also referred to as *Arabic number system*. (1)

decimal point: The point (.) in a decimal number that divides the whole part of the number from the fractional part of the number. For example, 12.75. (11)

degree: A unit of measure used for angles; 1/360 of a circle. (16)

denominate numbers: Numbers that represent measurements. For example, 17 feet is a denominate number, but the number 17 by itself is not a denominate number. (1)

denominator: The bottom number in a common fraction that refers to the total number of equal parts being considered in the fraction. (6)

derived units: Additional units of measure in SI that are arrived at through equations using the basic units. (24)

diameter: The straight line distance across a circle and passing through its center. (19)

difference: The answer that results when one number is subtracted from another. Also referred to as *remainder*. (3)

dividend: The number being divided in a division operation. (5)

division: The process of finding how many times one number (the divisor) can be contained in another number (the dividend). (5)

divisor: The number doing the dividing in a division operation. The dividend is divided by the divisor. (5)

double extra strong pipe: Pipe is generally classified as standard, extra strong, and double extra strong. The main difference among these is the wall thickness. For example, a standard 2" pipe has a wall thickness of 0.154", while a 2" double extra strong pipe has a wall thickness of 0.436". (2)

E

equilateral triangle: A triangle in which all sides are the same length and all angles are equal (60°). (18)

equivalent fraction: Fractions that are equal in value to each other but have different numerators and denominators from each other. For example, 1/2, 2/4, and 3/6 are equivalent fractions. The equivalency of any two fractions can be checked by reducing each fraction to its lowest terms. (6)

extra strong pipe: Pipe is generally classified as standard, extra strong, and double extra strong, the main difference being the wall thickness. For example, a standard 2" pipe has a wall thickness of 0.154" while a 2" extra strong pipe has a wall thickness of 0.218". (2)

F

fabrication: Construction of an object by putting parts together. (17)

filler material: The rod that a welder adds to the weld in order to increase the amount of material making up the joint. Also referred to as a *welding rod*. (7)

fillet weld: A weld with a triangular cross section.

flat-out length: The original length of a piece of metal that has been permanently bent, shaped, or formed. (22)

flux: A material used to prevent the formation of oxides during the brazing or soldering process. It is usually in a paste form and is rubbed on the filler rod and the joint. Its main ingredients are boric acids, chlorides, fluorides, and wetting agents. (5)

G

gage (GA): One of any numbering systems used to identify the various thicknesses of sheet steel. (3)

gallon: A standard American measurement of liquid capacity equal to 231 in³. The Canadian gallon contains 277.42 in³. (20)

galvanizing: The application of a thin coating of zinc to steel or iron in order to protect it from corrosion. In the commonly used hot dip method, the steel is dipped in molten zinc. See *cold galvanizing compound*. (14)

gear pump: A pump for delivering fluids. It causes a flow by passing the fluid between the teeth of two rapidly turning meshed gears. (14)

GMAW welder: The equipment (torch, wire, feed, power, gas flow, etc.) necessary for gas metal arc welding. This process uses a continuous consumable wire electrode that is fed through the torch. The weld itself is covered and protected by a shield of gas that is also fed through the torch. (4)

gram: A metric measurement of weight equal to 1/1000 kg. (21)

H

higher terms: A common fraction expressed as an equivalent fraction having higher values for numerator and denominator that could be reduced. (6)

hot-rolled steel: This refers to a manufacturing process used in making this steel. While the steel is still quite hot from manufacturing, it is passed through rollers to the required shape. Because it is being rolled while hot, this steel is rather dark and is not as accurately formed as cold-rolled steel. (10)

hypotenuse: The sloping line of a right triangle. It is the longest of the three sides. (18)

I

improper fraction: A common fraction with a numerator larger than the denominator, such as 7/6. (6)

International System of Units (SI). The modern form of the metric system, consisting of seven base units and their prefixes, which can be used to measure any quantity in the universe. (24)

invert the divisor: The process in the division of fractions in which the divisor is turned upside down to convert the division to multiplication. (10)

irregular object: An object having a cross section that is not uniform. (20)

isosceles triangle: A triangle in which two of the three sides are of equal length and two of the three angles are equal. (18)

K

kilogram (kg): A metric measurement of weight. It is equal to 1000 grams or approximately 2.205 lb. (21)

L

L shape: An abbreviation for *angle*. A structural steel shape with an *L* cross section. (21)

LCD. See *lowest common denominator*.

linear measure: Refers to measuring the straight line distance between two points. (15)

liter: A metric measurement of capacity. It is a little more than a quart (1.06 qt.) (20)

lower terms: A common fraction having numerator and denominator reduced to lesser values than those of equivalent fractions. (6)

lowest common denominator (LCD): A denominator for two or more fractions that is common to all the fractions and is the lowest number possible. (7)

lowest terms: A common fraction expressed as an equivalent fraction having the lowest possible values of numerator and denominator. (6)

M

mass: A measure of the amount of material in an object. (24)

meter (m): A metric measurement of length. It is equal to 1000 mm (or 100 centimeters) and is about 3" longer than a yard. (21)

metric system: A measuring system in which base units are used with prefixes to measure. Also known as *SI Metrics*. See *International System of Units* (SI). (24)

metric ton: A term commonly used for megagram, or 1000000 grams. (24)

millimeter (mm): The smallest metric measurement of length. It is 1/1000 of a meter or approximately 25.4". (15)

minuend: The larger of the two numbers in a subtraction operation. The subtrahend is subtracted from the minuend. (3)

minute: A unit of measure used for angles; 1/60 of a degree. (16)

mixed number: A number consisting of a whole number and a fraction. For example, 5 3/8. (6)

multiplicand: The number that is being multiplied in a multiplication operation. The multiplicand is multiplied by the multiplier. In other words, the multiplicand is being added to itself as many times as the value of the multiplier. (4)

multiplication: A fast and simplified method of adding that involves the repeated addition of a single number (the multiplicand) a set number of times (the multiplier). (4)

multiplier: The number that is doing the multiplying in a multiplication operation. The multiplicand is added to itself the number of times equivalent to the value of the multiplier. (4)

N

numerator: The top number in a common fraction, which refers to the number of parts out of the total number of equal parts being considered in the fraction. (6)

O

of: A word used to express the multiplication of fractions. For example, 1/5 of 3/4 means $1/5 \times 3/4$. (9)

oxyacetylene: A combination of oxygen and acetylene gas that, because it burns at an intense heat, is used as a fuel in welding and cutting. See *acetylene gas*. (13)

P

parallelogram: A shape having four sides. The sides opposite each other are the same length and are parallel. The angles opposite each other are equal. These are often described as looking like rectangles that have been tilted. (17)

payload: The maximum weight that a carrier (railroad car, airplane, etc.) can transport. (3)

percent (%): Parts of the whole of anything when the whole is divided into 100 equally sized parts. (23)

perimeter: The distance around a shape. (17)

pi (π): A Greek letter pronounced *pie*. It represents the number 3.14159 . . . . (19)

place value: The value of a number according to its place in a line of numbers. For example, the number 497 has three digits, each with a specific place value. Since the 7 is in the units digit, it has a value of 7. Since the 9 is in the tens digit, it has a value of 90. Since the 4 is in the hundreds digit, it has a value of 400. Each of these numbers in their place value combines to form the number 497. (1)

product: The final calculated value of a multiplication operation. (4)

proper fraction: A common fraction with the numerator smaller than the denominator. For example, 2/3. (6)

protractor: An instrument used to measure angles. (16)

Q

quotient: The answer arrived at in a division operation. It is the number of times the divisor fits into the dividend. (5)

R

radius: The straight line distance from the center of a circle to its edge. (19)

rectangle: A four-sided shape with four 90° angles. The opposite sides are parallel and equal in length. Adjacent sides are unequal in length. (17)

reducing: The process of changing improper fractions to mixed or whole numbers. Also, the process of expressing a common fraction in its lowest terms. (6)

regular object: An object having a uniform cross section. (20)

remainder: The amount remaining when a divisor will not divide evenly into a dividend. This term may also refer to the answer that results when one number is subtracted from another (the difference). (3, 5)

right triangle: A triangle in which one angle is 90°. (18)

root opening: The gap that may exist between two pieces to be welded. If the two pieces touch each other, the root opening is zero. (12)

rounding: The process of modifying accurate numbers into approximate numbers. After determining to which place value to round, evaluate the digit immediately to the right of that place value to determine whether to round up or down. If the value is 5 or higher, round up by 1. If below 5, leave the determined place value as is. Next, reduce down to zero all the digits to the right of the approximated digit. For example, to round the number 115,389 to the hundreds place, look to the number in the tens digit. Since it is an 8, round the number in the hundreds digit up by 1. Change the digits to the right of the hundreds digit to zero. The approximate number is 115,400. (1, 11)

rounding the decimal: The process of writing accurate or unending decimal numbers as approximate decimal numbers. For example, 15.69032 can be rounded to 15.69. (11)

S

S shape: An abbreviation for *standard shape*. A structural steel shape with a slope on the inside faces of the flanges of 1:6. (21)

second: A unit of measure used for angles; 1/60 of a minute. (16)

solder: The filler metal, usually an alloy or combination of tin and lead, that is used in soldering. The solder melts below 800°F (427°C) and joins the parts together. The weldment is heated, but not to the point of melting. (5)

square: A shape with four sides that are equal in length. The opposite sides are parallel and all angles in the square are 90°. (17)

square units: Units used in area measurement. Common units are square feet, square inches, and square millimeters. (17)

stainless steel: An alloy steel that has a strong resistance to corrosion because of the addition of chromium as an alloy. (7)

steel: Steel is an alloy of iron and carbon. To be classified as steel, the carbon content must be 1.7% or less. Steel that contains no other ingredients besides iron and carbon is called *plain carbon steel*. (1, 2, 3, 5, 6)

Steel Construction Manual: A book produced by the American Institute of Steel Construction that provides a full description of structural shapes. The Canadian equivalent is produced by the Canadian Institute of Steel Construction. (21)

subtraction: A process of "taking away" that determines how much larger one number is than another. (3)

subtrahend: In a subtraction operation, it is the number being subtracted. The subtrahend is subtracted from the minuend. (3)

sum: The final calculated number in an addition problem. (2)

surface grinder: A precision grinding machine on a pedestal. The workpiece is clamped in place on a movable worktable and then passed under a revolving grinding wheel. (14)

surfacing: A welding process whereby weld material is deposited on a surface in order to build it up. For example, the worn surface of a bulldozer blade may be built up by surfacing and then machined to proper dimension. (20)

T

terms: The numerator and denominator together are the terms of a fraction. (6)

3-4-5 triangle: A triangle with the length of the hypotenuse at 5, one side at 4, and the remaining side at 3, thus producing a right angle. (18)

tolerance: A specific range given of a dimension that a tradesperson can work within and still produce the desired results. (15)

trapezoid: A four-sided shape in which two sides are parallel and the other two sides are not parallel. (17)

triangle: A shape having three straight sides. (18)

U

US customary system: A common measuring system using feet and inches. For example, 12 inches make up one foot. (15)

US standard gage: One of the most popular gages used in classifying sheet steel thicknesses. (8)

V

V-groove joint: A type of joint in which two flat pieces are beveled on their sides to form a V-shape. (5)

volume: The amount of space an object occupies. (20)

W

W shape: An abbreviation for *wide flange shape*. This steel shape is characterized by a constant thickness of the flange. In general, it has a greater flange width and a relatively thinner web than S shapes. (21)

weight: The gravitational force exerted by Earth (or another celestial body) on an object. (24)

weldment: Any metal object made primarily by welding. (2, 3, 4)

Index

Answers to Odd-Numbered Practice Problems

Section 1: Whole Numbers

Unit 1: Introduction to Whole Numbers, page 5

1.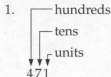
   hundreds
   tens
   units
   471

3.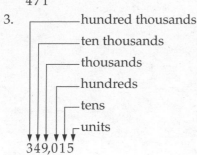
   hundred thousands
   ten thousands
   thousands
   hundreds
   tens
   units
   349,015

5. 694,700
7. 17,200
9. 29,900
11. 946,000
13. 11,000
15. 449,000
17. 1,295,000
19. 13,001,000
21. A. 250 psi, B. 11,500 ft²,
   C. five acres, D. 17.5 tons,
   G. $7.65, H. $7.65 per hour,
   I. 40" per minute, K. 12 dozen

Unit 2: Addition of Whole Numbers, page 11

1. 706
3. 34,472
5. 89,454
7. 37
9. 825,833
11. 1,140
13. 14,000
15. 4,333
17. 1,141,120
19. 88"
21. 366,433 welding rods
23. 159 hours
25. 1,766 women
27. 541"
29. 552"
31. 209"
33. 9,028 hours

Unit 3: Subtraction of Whole Numbers, page 21

1. 664
3. 1,746
5. 20,019
7. 42,630
9. 1,015,423
11. 5,722
13. 7,222
15. 27
17. 93,640
19. 805 brackets
21. 34,730 lb of 16 gage
23. 20,831 lb
25. 1,980 of #51 tips
27. 413 lb
29. 32,835 lb of steel
31. 191"
33. 91"

35. 158" of channel
37. 406" of flat bar

Unit 4: Multiplication of Whole Numbers, page 35

1. 408
3. 3,920
5. 174,915
7. 4,748,576
9. 1,876,070
11. 4,037,688
13. 2,508,156
15. 4,680
17. 637,512
19. 18,914 spot welds
21. 1,053 studs
23. 52,254 holes drilled
25. 7,948 lb
27. 63,138' of 1 1/2' pipe
29. 238,954' of pipe
31. 42 lb
33. 26,588"
35. 765 lb
37. 3,500 lb

Unit 5: Division of Whole Numbers, page 47

1. 27
3. 103
5. 9,007 r 1
7. 7
9. 400
11. 25 r 40
13. 64 r 44
15. 25 r 70

17. 19 r 4
19. 61 lb
21. 6 pieces
23. 383,916 lb
25. 183′
27. 8″
29. 6″
31. 985 pieces
33. 252 studs

Section 2: Common Fractions

Unit 6: Introduction to Common Fractions, page 57

1. 10/16
3. 98/98
5. 7/8
7. 55/65
9. 80/100
11. yes
13. no
15. yes
17. yes
19. 41/2
21. 175/8
23. 31/2
25. 101/30
27. 2
29. 51/4
31. 233/8
33. 1670/9
35. 110/10 (or reduced to 11/1)
37. 1/2
39. 7/8
41. 4/41
43. 3/5
45. 101/200

Unit 7: Addition of Fractions, page 63

1. 11/18
3. 168/219
5. 2 23/50
7. 2 23/60
9. 1 11/24
11. 1 2,451/16,192
13. 23 7/16
15. 158 97/204

17. 53/238
19. 402 1/4 hours
21. 65 11/16″
23. 22 9/16″
25. 41 61/64″
27. 79 45/64″
29. 92 15/32″
31. 73 7/32″
33. 505 55/64″

Unit 8: Subtraction of Fractions, page 71

1. 1/3
3. 31/101
5. 1/16
7. 689 15/16
9. 111/238
11. 15 1/3
13. 2/3
15. 7/15
17. 12 2/3
19. 62 13/16″
21. 82 3/8 miles
23. 72 1/8 miles
25. 2 7/8″
27. 66 45/64″
29. 9 5/16″
31. 79/16″
33. Length B: 55/8″
35. Dimension A: 19 13/32″
37. 3015/16″
39. 1011/32″
41. 57/32″

Unit 9: Multiplication of Fractions, page 81

1. 1/15
3. 60/671
5. 13/72
7. 3/4
9. 39 3/20
11. 87 57/64
13. 20/63
15. 2511/64
17. 25/8
19. 3 11/16 miles
21. 200 1/4 hours
23. 71 1/2′
25. 10,802,500 bolts

27. 237 7/16″
29. 1,125 3/16 lb
31. 270 rods
33. 3,319 1/32 lb
35. 378 lb
37. 105,252″
39. 68,316 3/4″

Unit 10: Division of Fractions, page 91

1. 11/15
3. 20/21
5. 9/16
7. 25 1/5
9. 21 5/7
11. 2 33/57
13. 29/36
15. 1 9/112
17. 5/12
19. 8 pieces
21. 35 miles per hour
23. 49 pieces
25. 236 13/15 ft^3
27. 175/8″
29. 10 pieces
31. 131 9/64″

Section 3: Decimal Numbers

Unit 11: Introduction to Decimal Numbers, page 101

1. 100.01
3. 14.00125
5. 0.707
7. 0.866
9. 9/20
11. 267/2,000
13. 3,333/10,000
15. 0.5
17. 0.03125
19. 0.625
21. 0.9
23. 0.5
25. 0.1
27. 0.96
29. 0.09
31. 0.91
33. 0.314

35. 0.009
37. 0.443
39. 0.67
41. 0.11
43. 0.10
45. 0.938
47. 0.063
49. 0.767
51. 0.0156
53. 0.1235
55. 0.9659

Unit 12: Addition and Subtraction of Decimal Numbers, page 105

1. 13.4
3. 363.398
5. 52.66532
7. 222.49
9. 104.3088
11. 0.7
13. 17.6982
15. 4.093
17. 5.0644
19. 345.4892
21. 1.757"
23. 88.328"
25. + $97.55
27. Rod B: 1.413"
29. Rod D: 2.7198"
31. Rod F: 4.0266"
33. Rod H: 5.3334"
35. 30.3588"
37. Block A: 1.45625"
39. Block C: 1.24585"
41. 15.8724"

Unit 13: Multiplication of Decimal Numbers, page 115

1. 0.3026
3. 70.356
5. 1,001.01001
7. 6.66
9. 1,093.75
11. 29.9832
13. 0.00008
15. 1.13918
17. 195,000
19. 65.20293
21. 1.367631

23. 68,200 lb
25. $22.40
27. 51.75 ft³
29. 83,145.258 lb
31. $240.60
33. 3,504.879"
35. 112.271"
37. 2,038.06528 lb
39. 1,872.54846 lb

Unit 14: Division of Decimal Numbers, page 125

1. 3,500
3. 0.197
5. 8
7. 0.07
9. 1.00
11. 136.96
13. 3.000
15. 0.017
17. 0.719
19. 16 hours
21. 750 ball bearings
23. 4 blocks
25. 623 items
27. 364 squares
29. 54 applications
31. 5.6233"

Section 4: Measurement

Unit 15: Linear Measure, page 141

1. 108 mm, 4 1/4"
3. 74 mm, 2 15/16"
5. 18 mm, 23/32"
7–12. Individual answers will vary.
13. 423"
15. 544.5"
17. 6,042"
19. 63,360"
21. 3 3/16"
23. 145"
25. 9'
27. 400'-1"
29. 17'-11 3/4"
31. 65'-7.96"
33. 11/32'
35. 17/768'

37. 304.8 mm
39. 254.0 mm
41. 0.4 mm
43. 15.9 mm
45. 1605.0 mm
47. 209.6 mm
49. 39.370"
51. 0.039"
53. 2.579"
55. 466.535"
57. 7.677"
59. 207.874"
61. 15 3/4"
63. 48"
65. 19 3/8"
67. 19 14/16"
69. 1"
71. 21 1/16"
73. 11 21/32"
75. 730 6/32"
77. 31/32"
79. 68 2/64"
81. 12 35/64"
83. 632 26/64"
85. 14'-8 7/8"
87. 362.1875 ft²; 52,155 in²
89. 13'-3/8"
91. 249 mm; 247 mm
93. 1.345"; 1.335"
95. 15 mm; 14 mm
97. 20.1 mm; 19.7 mm
99. 3 90/100"; 3 93/100"; 3 87/100"
101. 87 60/100"; 87 63/100"; 87 57/100"
103. 7/16"; 5/16"
105. 8 13/16"; 8 11/16"
107. 15 5/16"; 15 3/16"
109. A. 4 7/16"
     B. 2 3/4"
     C. 3 1/8"
     D. 1 1/4"
     E. 1 7/8"
     F. 7/16"
111. Check answers individually.
113. A. 4 mm
     B. 27 mm
     C. 65 mm
     D. 110 mm

115. 0.438"; 11.13 mm
117. 0.500"; 12.70 mm
119. 0.531"; 13.49 mm
121. 4.969"; 126.21 mm
123. 37.5 mm; 37.0 mm
125. 149.5 mm; 149.0 mm
127. 160.1 mm; 159.6 mm
129. 5 mm; 0.20"; 13/64"
131. 78 mm; 3.07"; 3 4/64"
133. 13 mm; 0.51"; 33/64"
135. 50 mm; 1.97"; 1 62/64"

Unit 16: Angular Measure, page 155

1. 90°
3. 20°
5. 215°
7. 10° 38'
9. 75° 21'
11. 200° 39' 23"
13. 66° 50' 55"
15. 28° 2' 20"
17. 33° 59' 2"
19. 293°43' 42"
21. 11° 52' 28"
23. 283° 14' 56"
25. 909° 50' 12"
27. 1,239° 16' 42"
29. 131° 14' 9"
31. 30°
33. 51° 25' 43"
35. 8° 27' 59"
37. 90°; 85°
39. 40° 35'; 33° 25'
41. 30° 30"; 29° 59' 30"
43. 212° 47'; 212° 17'
45.

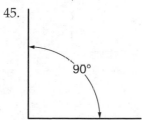

47.

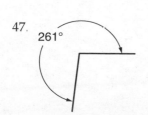

49.

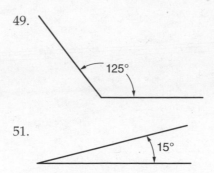

51.

53. 30°
55. 20°

Unit 17: Four-Sided Measure, page 167

1. 5,040 in²
3. 1,400 in²
5. 65 in²
7. 1 ft²
9. 111 ft²
11. 694 4/10 ft²
13. 58064 mm²
15. 88468 mm²
17. 645 mm²
19. 2.4 in²
21. 0.2 in²
23. 1.0 in²
25. 65,985 in²
27. 252537 mm²
29. 9,919,872 in²
31. 2,351.25 in²
33. 39354720 mm²
35. 13222 mm²
37. 347 1/2"
39. 6,217 1/2 in²

Unit 18: Triangular Measure, page 177

1. 12000 mm
3. 161'
5. 336"
7. 698.705 in²
9. 5565174.8 mm²
11. 2,812.5 ft²
13. 82"
15. 4,896 in²
17. 332.544'
19. 556.6'
21. 23500 mm

Unit 19: Circular Measure, page 191

1. 2,582 in²
3. 166'-5"
5. 12 14/16 in²
7. 2,499 in²
9. 4,068 in²
11. 576334 mm²
13. 395.3"
15. 156000 mm
17. 450"

Section 5: Volume, Weight, and Bending Metal

Unit 20: Volume Measure, page 205

1. 1,443.75 in³
3. 11,968 gallons
5. 127 gallons
7. 79 ft³
9. 13,824 in³
11. 737415 mm³
13. 269.28 gallons
15. 155 in³
17. 176839.27 mm³
19. 11 3/4 yd³
21. 8,117.34 in³
23. 1200000 mm³
25. 15,660 in³
27. 4 ft³
29. 281893.5 mm³
31. The rectangular container
33. 926 1/4 in³
35. 42,390 in³
37. 813750 mm³

Unit 21: Weight Measure, page 221

1. 15 kg
3. 949 lb
5. 214.5 kg
7. 370 lb
9. 16.73 lb
11. 1,706.4 lb
13. 7,706 lb
15. 227 lb
17. 10635 grams
19. 181.3 lb

**Unit 22: Bending Metal,
page 239**

1. 478 1/8″
3. 20.625″ × 23″
5. 560 mm
7. 208 1/32″
9. 43 41/64″ × 50″
11. 163 15/64″
13. 87 3/4″
15. 1257 mm

Section 6: Percentages and the Metric System

Unit 23: Percentages, page 251

1. 13/100
3. 1/2,000
5. 10 1/10
7. 0.75
9. 0.1414
11. 0.00007
13. 75%
15. 3333 1/3%

17. 57 17/19%
19. 12.5%
21. 1.9%
23. 0.3%
25. 288.05
27. 28.5
29. 0.21
31. 22.46%
33. 50%
35. 900%
37. 207 parts per hour
39. 73%
41. $411,450,000
43. $514.95, $92.69, $18.02, $20.60
45. $335.20, $50.28, $11.73, $13.41
47. $421.10, $75.80, $14.74, $16.84
49. $417.74, $71.02, $14.62, $16.71

**Unit 24: The Metric System,
page 265**

1. meter (m), kilogram (kg), second (s), ampere (A), kelvin (K), candela (cd), and mole (mol)

3. Meter
5. 100 cm
7. 15.97 m
9. 2985.4 km
11. 0.0595 km
13. 914 mm
15. 419 mm
17. 91.44 m
19. 3 29/64″
21. 3,280′-10 4/64″
23. 32,808′-4 50/64″ (or reduced to 32,808′-4 25/32″)
25. 1000 ml
27. 0.182 L
29. 6700 ml
31. 2.64 gallons
33. 1000 g
35. 39250 g
37. 0.212 g
39. 0.065 kg
41. 14.27 lb
43. 3175 kg
45. 62597 kg